Springer Theses

Recognizing Outstanding Ph.D. Research

Aims and Scope

The series "Springer Theses" brings together a selection of the very best Ph.D. theses from around the world and across the physical sciences. Nominated and endorsed by two recognized specialists, each published volume has been selected for its scientific excellence and the high impact of its contents for the pertinent field of research. For greater accessibility to non-specialists, the published versions include an extended introduction, as well as a foreword by the student's supervisor explaining the special relevance of the work for the field. As a whole, the series will provide a valuable resource both for newcomers to the research fields described, and for other scientists seeking detailed background information on special questions. Finally, it provides an accredited documentation of the valuable contributions made by today's younger generation of scientists.

Theses are accepted into the series by invited nomination only and must fulfill all of the following criteria

- They must be written in good English.
- The topic should fall within the confines of Chemistry, Physics, Earth Sciences, Engineering and related interdisciplinary fields such as Materials, Nanoscience, Chemical Engineering, Complex Systems and Biophysics.
- The work reported in the thesis must represent a significant scientific advance.
- If the thesis includes previously published material, permission to reproduce this must be gained from the respective copyright holder.
- They must have been examined and passed during the 12 months prior to nomination.
- Each thesis should include a foreword by the supervisor outlining the significance of its content.
- The theses should have a clearly defined structure including an introduction accessible to scientists not expert in that particular field.

More information about this series at http://www.springer.com/series/8790

Ulrich Römer

Numerical Approximation of the Magnetoquasistatic Model with Uncertainties

Applications in Magnet Design

Doctoral Thesis accepted by
Technische Universität Darmstadt, Germany

Author
Dr.-Ing Ulrich Römer
Institut für Theorie Elektromagnetischer Felder
Technische Universität Darmstadt
Darmstadt
Germany

Supervisor
Prof. Thomas Weiland
Institut für Theorie Elektromagnetischer Felder
Technische Universität Darmstadt
Darmstadt
Germany

ISSN 2190-5053 ISSN 2190-5061 (electronic)
Springer Theses
ISBN 978-3-319-82316-4 ISBN 978-3-319-41294-8 (eBook)
DOI 10.1007/978-3-319-41294-8

Printed on acid-free paper

This Springer imprint is published by Springer Nature
The registered company is Springer International Publishing AG Switzerland

Supervisor's Foreword

Uncertainties and manufacturing tolerances play an important role in accelerator physics, in view of extreme design requirements and high production costs. Quantifying uncertainties in magnetic fields is paramount for the design of accelerator magnets and other engineering applications such as machines and transformers. This thesis proposes a thorough and general mathematical approach for modeling and solving stochastic magnetic field problems. Variability in material properties and the geometry is discussed with emphasis on the preservation of structural physical and mathematical properties. The latter is used to show the unique solvability of the resulting nonlinear stochastic equations. Also, efficient discretization methods for random inputs are described, which are new in this context. The stochastic partial differential equations are solved using adjoint based perturbation methods as well as recently emerged higher order spectral approximation schemes. Thereby, this thesis gives both theoretical results on convergence rates and error estimation and addresses numerical examples with practical relevance, in particular recently designed accelerator magnets.

Darmstadt, Germany
December 2015

Prof. Thomas Weiland

Parts of this thesis have been published in the following documents:

Journal articles

[1] Andreas Bartel, Herbert De Gersem, Timo Hülsmann, Ulrich Römer, Sebastian Schöps, and Thomas Weiland. Quantification of uncertainty in the field quality of magnets originating from material measurements. *IEEE Transactions on Magnetics*, 49(5):2367–2370, 2013.
[2] Ulrich Römer, Sebastian Schöps, and Thomas Weiland. Approximation of moments for the nonlinear magnetoquasistatic problem with material uncertainties. *IEEE Transactions on Magnetics*, 50(2), 2014.
[3] Ulrich Römer, Sebastian Schöps, and Thomas Weiland. Stochastic modeling and regularity of the nonlinear elliptic curl-curl equation. SIAM/ASA Journal on Uncertainty Quantification (in press).

Acknowledgments

I am very grateful for the support and encouragement that I received during my time as a Ph.D. student. In particular I would like to thank:

- Prof. Thomas Weiland for the supervision and support and for providing and excellent working atmosphere at the Institut für Theorie Elektromagnetischer Felder
- Prof. Stefan Ulbrich for kindly accepting to referee the thesis and for very valuable comments and discussions on nonlinear partial differential equations
- Prof. Sebastian Schöps for the supervision and support and for numerous valuable discussions on uncertainties and numerics
- Dr.-Ing. Stephan Koch for the supervision at the beginning of the thesis
- My (former) colleagues at TEMF, in particular, Jens Trommler, Martin Lilienthal, Klaus Klopfer and Cong Liu
- Uwe Niedermayer, Prof. Johann Baumeister and Prof. Peter Kloeden for comments on the thesis
- Michaela Arnold for the very good cooperation on magnet design
- My family and friends

I would also like to acknowledge the support of the Deutsche Forschungsgemeinschaft through CRC 634.

Contents

Notations

a.e.	If a condition holds on a set V, modified by a subset of measure zero, the condition is said to hold almost everywhere (a.e.) on V, see Ref. [3, p. 15] of Chap. 3. In particular in $L^P(V)$, $1 \leq p \leq \infty$, functions are identified a.e. on V. 18
a.s.	An event that occurs with probability one is said to occur almost surely (a.s.). 66
f_{BH}	Relation between absolute values of magnetic flux density and field, called B–H curve. 10
f_{HB}	Relation between absolute values of magnetic field and flux density, inverse function of B–H curve. 10
$P_h^{\mathbf{grad}}, P_h^{\mathbf{curl}}, P_h^{\mathrm{div}}, P_h$	Projection operators from the spaces $\mathcal{H}(\mathbf{grad}, D), \mathcal{H}(\mathbf{curl}, D), \mathcal{H}(\mathrm{div}, D), \mathrm{L}^2(D)$, into their discrete counterparts. 23
χ_V	Characteristic function of the set V. 44
$\mathbf{H}_{\mathrm{CO}}$	Coercive magnetic field strength. 29
$\mathbf{J}_{\mathrm{CO}}$	Current associated to coercive magnetic field through $\mathbf{J}_{\mathrm{CO}} = \mathbf{curl}\,\mathbf{H}_{\mathrm{CO}}$. 30
$\mathbf{V}$	Covariance matrix. 80
$\mathcal{U}$	Curl-harmonic extension operator. 18
$\mathrm{curl}_S\, \mathbf{u}$	Scalar curl operator at a surface, $\mathrm{curl}_S\, u = \mathbf{curl}\, \mathbf{u} \cdot \mathbf{n}$. 55
$\mathbf{curl}_S\, u$	Vectorial curl operator at a surface, $\mathbf{curl}_S\, u = \mathbf{grad}\, u \times \mathbf{n}$, an equivalent definition is given by $\mathbf{curl}_S\, u = \mathbf{grad}_S\, u \times \mathbf{n}$, with the tangential gradient. 55
$\mathrm{d}\mathbf{A}$	Oriented surface element, also denoted $\mathrm{d}\mathbf{x}$. 5
$\mathrm{d}\mathbf{s}$	Oriented line element, also denoted $\mathrm{d}\mathbf{x}$. 5
$\mathrm{d}V$	Volume element, also denoted $\mathrm{d}\mathbf{x}$. 5
δ_g	Fréchet derivative of $g: U \to Y$, referring to the continuous linear operator $g: X \to Y$ such that $g(u+v) = g(u) + \delta g(v) + o(\| v \|)$. 21

g'	Gâteaux derivative of $g : X \to Y$, referring to the continuous linear operator $g' : X \to Y$ such that $\lim_{s \to 0}(g(u+sv) - g(u))/s = g'(v)$. 21
$\det(D\mathbf{F})$	Determinant of matrix $D\mathbf{F}$. 25
diam	Diameter of a subset of a metric space. 8
$\mathcal{U}(-\sqrt{3}, \sqrt{3})$	Uniform distribution on the interval $(-\sqrt{3}, \sqrt{3})$. 69
D_{obs}	Dobs observer region $D_{\text{obs}} \subset D_{\text{E}}$, where the quantity of interest is evaluated. 40
D_{C}	Domain with homogeneous conductivity and nonlinear material law. 7
D_{E}	Air part of the computational domain. 7
D_{HA}	Support of domain perturbations referred to as hold-all. 43
$D_{\mathbf{J}}$	Domain of the imposed current density. 7
E	Magnetic energy. 21
$\mathbf{H}_{\text{eq}}$	Equilibrated flux used within a Prager-Synge identity. 33
err_h	Error measure for discretization error. 32
err_{L}	Error measure for linearization error. 32
$\text{err}_{\text{L},h}$	Error measure for linearization and discretization error for nonlinear problems. 32
$\lvert\cdot\rvert$	Euclidean norm of a vector. 7
$\boldsymbol{A}_{h,h_T}$	Approximation of $\boldsymbol{A}$ by means of finite elements in physical space and the implicit Euler scheme in time. 36
$\mathbf{A}_{\text{L}}$	Field approximated by means of a linearization procedure. 28
$\mathbf{A}_{\text{L},h}$	Field approximated by means of a linearization and finite element method. 28
$\mathbf{A}_{\text{L},0}$	Linearization point at any step of the linearization-iteration procedure. 28
$\mathbf{A}_q$	Collocation approximation of $\mathbf{A}$. 85
$\mathbf{A}_{h,q}$	Collocation approximation of $\mathbf{A}_h$. 86, 112
$\mathbf{A}_h$	Finite element approximation of $\mathbf{A}$. 26
$\mathbf{A}_{\text{init}}$	Initial condition for $\mathbf{A}$. 18
$\mathbf{A}^{\{l\}}$	l-th iterate within linearization scheme. 107
$\mathbf{A}_{\text{C}}$	Restriction of $\mathbf{A}$ to D_{C}. 12
$\mathbf{A}_{\text{E}}$	Restriction of $\mathbf{A}$ to D_{E}. 12
$\mathbf{a}_n$	Approximation at n-th time step within the implicit Euler scheme. 36
$\nu^{(1)}(\cdot, s)$	Partial derivative with respect to the second argument, i.e., $\nu^{(1)}(\cdot, s) = \partial_s \nu(\cdot, s)$. 22
$a(\mathbf{w}; \cdot, \cdot)'$	Bilinear form, linearization of a at $\mathbf{w}$. 23
$l(\mathbf{v})$	Linear form on $L^2(D)^3, l(\mathbf{v}) := (\mathbf{J}, \mathbf{v})_D$. 19
$a(\mathbf{A}; \mathbf{v})$	Form $a(\mathbf{A}; \mathbf{v}) := (\mathbf{h}(\cdot, \mathbf{curl}\,\mathbf{A}), \mathbf{curl}\,\mathbf{v})_D$, nonlinear and linear in the first and second argument, respectively. 19
Γ_{D}	Dirichlet boundary of the computational domain. 9

Γ_{I}	Interface between conducting and non-conducting domain. 9
$\hat{Q}_i(Y_i)$	Univariate term in the HDMR expansion. 75
$\hat{Q}_{ij}(Y_i, Y_j)$	Bivariate term in the HDMR expansion. 75
$\hat{Q}_0$	Constant term in the HDMR expansion. 75
B_i^{ms}	Flux density values for measurements. 42
H_i^{ms}	Measured magnetic field strength at B_i^{ms}. 42
$\mathbf{I}$	3×3 identity matrix. 22
$\boldsymbol{\beta}$	Generic, possibly infinite-dimensional, input parameter representing shape, material coefficient our source current. 39
$\tilde{\boldsymbol{\beta}}$	Perturbation of the input parameter. 40
$D\mathbf{h}$	Jacobian of vector function $\mathbf{h}$, $D\mathbf{h}(\cdot, \mathbf{r})_{ij} = \left(\partial h_i(\cdot, \mathbf{r})/\partial r_j\right)_{ij}$. 22
D^2	Second variation. 87
$[\![\mathbf{A}]\!]_S$	Tangential jump of the vector $\mathbf{A}$ at the face S, $[\![\mathbf{A}]\!]_S := \mathbf{A}^+ \times \mathbf{n}^+ + \mathbf{A}^- \times \mathbf{n}^-$. 11
$[\mathbf{u}]_S$	Difference of the vector or scalar $\mathbf{u}$ at the face S, $[\mathbf{u}]_S := \mathbf{u}^+ - \mathbf{u}^-$. 55
$\mathcal{O}$	Landau symbol, for two functions f, g, $f(x) = \mathcal{O}(g(x))$ if $\lim_{x\to a} \frac{f(x)}{g(x)} = C$. 7
o	Landau symbol, for two functions f, g, $f(x) = o(g(x))$ if $\lim_{x\to a} \frac{f(x)}{g(x)} = 0$. 21
$\mathbf{h}_{\mathrm{L}}$	Linearized vector function by means of the Picard or Newton–Raphson method. 28
B	Magnitude of the magnetic flux density $B := \lvert\mathbf{B}\rvert$. 42
H	Magnitude of the magnetic field $H := \lvert\mathbf{H}\rvert$. 42
$\mathbf{O}$	Matrix with all zero entries. 26
$\mathcal{T}_h$	Quasi-uniform family of subdivision of the computational domain. 23
$\mathcal{T}_T$	Quasi-uniform subdivision of the temporal domain. 35
min!	$f(x) = \min!$, $x \in X$, refers to $\min_{x\in X} f(x) = \alpha$. 21
F	Functional for the variational formulation of magnetostatic system $F(\mathbf{u}) := E(\mathbf{u}) - l(\mathbf{u})$. 21
$\mathrm{Cov}[g]$	Covariance of g. 67
$\tilde{\mathrm{M}}^k[\mathrm{g}]$	k-th statistical moment of g. 77
M^k	k-th centered statistical moment. 77
$\mathrm{E}[\mathrm{g}]$	Expected value of g. 67
Std	Standard deviation. 76
$\mathrm{Var}[\mathrm{g}]$	Variance of g. 77
A_n	Skew multipole coefficients. 93
$\bar{A}_n(r_0)$	Integrated skew multipole coefficients. 94

B_n	Normal multipole coefficients. 94
$\bar{B}_n(r_0)$	Integrated normal multipole coefficients. 94
N_{cp}	Number of control points. 45
N_{e}	Number of edges in the finite element mesh. 35
N^{MC}	Number of Monte Carlo samples. 76
N^{ms}	Number of measurement points. 42
N_{N}	Number of nodes in the finite element mesh. 26
N^q	Number of collocation points. 84
N_{str}	Number of strands in the stranded conductor model. 47
N_T	Number of time steps. 35
Ω	Set of random outcomes. 66
$\mathcal{M}$	Nonlinear operator associated to the stiffness term. 20
$\mathbf{A}[\beta]$	Dependence of the solution $\mathbf{A}$ on a parameter β. 39
$\mathbf{y}$	Generic real, deterministic and finite dimensional parameter vector. 41
Γ	Range of parameter vector $\mathbf{y}$. 65
∂_s	Partial derivative ∂/∂_s. 10
$\mathbf{F}^*$	Pullback of transformation $\mathbf{F}$, i.e., see Ref. [18, p. 8] of Chap. 2 for a definition. 25
P	Measure defined on $\mathcal{F}$. 66
$\mathcal{P}_q(K)$	The space of polynomials up to degree q on $K \subset \mathbb{R}^3$. 24
$S_q(K)$	Polynomial space $S_q(K) = \{\mathbf{s} \in \mathbb{P}_q(K)^3 \vert \mathbf{s}(\mathbf{x}) \cdot \mathbf{x} = 0, \forall \mathbf{x} \in K\}$. 24
$\mathcal{S}_N^{q,k}$	B-spline space span $\left\{B_j^q\right\}_{j=1}^N$ of polynomials of degree q on sub-intervals of $\prod$ and regularity k at the knots. 25
$\mathbb{Q}_q$	The space of tensor product polynomials of degree at most q. 84
$\mathcal{I}_{q_n}$	Lagrange interpolation operator in dimension n of degree q_n. 84
Ψ_M	Sum of first M eigenvalues divided by all eigenvalues. 68
Q	Quantity of interest, such as multipole coefficients, the inductance or a specific norm. 40
$\mathbb{R}$	Set of real numbers. 7
ν	Magnetic reluctivity, the inverse permeability. 10
ν_{d}	Differential reluctivity. 22
$\boldsymbol{\nu}_{\mathrm{d}}$	Differential reluctivity tensor. 22
$\mathbb{R}_0^+$	Set of positive real numbers with zero. 9
A	Bounded, piecewise smooth and oriented surface, if not stated otherwise. 5
U_{adm}	Abstract set of admissibility for the input parameters. 39
$\tilde{U}$	Space of perturbations. 40
$U_{\mathrm{adm}}^{\mathrm{I}}$	Set of admissible interfaces. 43
$U_{\mathrm{adm}}^{\mathbf{J}}$	Admissible set for the source current. 47

U^{ν}_{adm}	Set of admissibility for the magnetic reluctivity. 41
S	Bounded, piecewise smooth and oriented line, if not stated otherwise. 6
V	Bounded, piecewise smooth and oriented volume, if not stated otherwise. 5
$\mathcal{T}_s$	Domain deformation mapping. 43
$\mathcal{V}$	Velocity field used in the velocity method to model deformed domains. 43
$\mathcal{F}$	Sigma algebra associated to Ω. 66
$\mathcal{Z}$	Abstract implicit state operator for the parametrized model equations. 39
supp	Support of a function. 7
$\dot{\mathbf{A}}$	Derivative of the (Banach space valued) function $\mathbf{A}(t)$. 19
$\otimes$	Tensor product. 22
I_T	Time interval $I_T = (0, T]$. 11
$\mathbf{x}$	Arbitrary point of free space $\mathbb{R}^3$. 7
$\mathbf{X}$	Lagrangian variable. 43
∂V	Boundary of V. 5
$\mathbf{h}$	Abstract vector function $\mathbf{h}(\mathbf{x}, \mathbf{r}) := \nu(\mathbf{x}, \lvert\mathbf{r}\rvert)\mathbf{r}$, representing the magnetic field. 19
wcs	Worst case scenario for a quantity of interest. 86
wcs_{L}	Approximate linear worst case scenario for a quantity of interest. 87
$\boldsymbol{\chi}$	Winding function for conductor models. 47
δ_{ij}	Kronecker delta $\delta_{ij} = 1$ if $i = j$ and $\delta_{ij} = 0$ if $i \neq j$. 22

Function Spaces

$\mathcal{H}(\mathbf{curl}, D)^*$	Dual of the space $\mathcal{H}(\mathbf{curl}, D)$. 20
$\mathcal{C}(X, Y)$	Space of continuous functions from X into Y. 19
$\mathcal{C}^{k,1}(\bar{V})$	The space of k-times differentiable functions, that are Lipschitz continuous on V, where $k \in \mathbb{N}_0$. 43
$\mathcal{C}^k(\overline{\mathbb{R}^+})$	Functions in $\mathcal{C}^k(\mathbb{R}^+)$ with bounded and uniformly continuous derivatives. 41
$\mathcal{C}_0^k(\overline{\mathbb{R}^+})$	Functions in $\mathcal{C}^k(\overline{\mathbb{R}^+})$ with compact support. 41
$\Vert u\Vert_{\mathcal{C}^k(\overline{\mathbb{R}^+})}$	Norm on the space $\mathcal{C}^k(\overline{\mathbb{R}^+})$, defined as $\Vert u\Vert_{\mathcal{C}^k(\overline{\mathbb{R}^+})} := \max_{1\leq i\leq k} \sup_{s\in\mathbb{R}^+} \left\vert\partial_s^i u(s)\right\vert$. 41
$G(V)$	Gradients of functions in $\mathcal{H}(\mathbf{grad}, V)$ that are constant on each connected part of the boundary. 18
$< \mathcal{M}\,\mathbf{A}, \mathrm{v} >$	Duality pairing of $\mathcal{M}\,\mathbf{A} \in \mathcal{H}(\mathbf{curl},\, D)^*$ and $\mathrm{v} \in \mathcal{H}(\mathbf{curl}, D)$. 20
$\mathcal{H}(\mathbf{curl}, D)$	$\mathbf{u} \in L^2(D)^3$ Such that there exists $\mathbf{w} \in L^2\,(D)^3$ and $(\mathbf{w},\mathbf{v})_D = (\mathbf{u}, \mathbf{curl}\,\mathbf{v})_D$, for all $\mathbf{v} \in \mathcal{C}_0^\infty(D)^3$. 19
$\Vert\mathbf{u}\Vert_{\mathcal{H}(\mathbf{curl},D)}$	Norm of the space $\mathcal{H}(\mathbf{curl}, D)$, $\Vert\mathbf{u}\Vert_{\mathcal{H}(\mathbf{curl},D)} := \Vert\mathbf{curl}\,\mathbf{u}\Vert_2$. 19
$\mathcal{H}(\mathrm{div}, D)$	$\mathbf{u} \in L^2(D)^3$ Such that their exists $\mathbf{w} \in L^2(D)^3$ such that $(\mathbf{w},\mathrm{v})_D = -(\mathbf{u}, \mathbf{grad}\,\mathrm{v})_D$, for all $\mathrm{v} \in \mathcal{C}_0^\infty(D)$. 24
$\mathcal{H}(\mathrm{div}\ 0, D)$	The space of square-integrable functions $\mathbf{u}$ on D, such that $(\mathbf{u}, \mathbf{grad}\,\mathrm{v})_D = 0$, for $\mathrm{v} \in \mathcal{C}_0^\infty(D)$. 47
$\mathcal{H}(\mathbf{grad}, D)$	$u \in L^2(D)$ Such that their exists $\mathbf{w} \in L^2(D)^3$ and $(\mathbf{w}, \mathbf{v})_D = -(u, \mathrm{div}\,\mathbf{v})_D$, for all $\mathbf{v} \in \mathcal{C}_0^\infty(D)^3$. 17
$\mathcal{H}_h(\mathbf{curl}, D)$	Finite dimensional subspace of $\mathcal{H}(\mathbf{curl}, D)$. 24
$\mathcal{H}_h(\mathrm{div}, D)$	Finite dimensional subspace of $\mathcal{H}(\mathrm{div}, D)$. 24
$\mathcal{H}_h(\mathbf{grad}, D)$	Finite dimensional subspace of $\mathcal{H}(\mathbf{grad}, D)$. 24
$L_h^2(D)$	Finite dimensional subspace of $L^2(D)$. 24
$\mathcal{H}^s(D)^3$	Fractional order Sobolev space with non-integer s. 28

$\mathcal{H}^s(\mathbf{curl}, D)$	Functions $\mathbf{u} \in \mathcal{H}^s(D)^3$ such that the (weak) curl satisfies $\mathbf{curl}\,\mathbf{u} \in \mathcal{H}^s(D)^3$ with non-integer s. 28
$\|\mathbf{u}\|_{L^\infty(D)}$	Uniform norm of the space $\|\mathbf{u}\|_{L\infty(D)} := \text{ess sup}_{x\in D}\|\mathbf{u}\|$. 33
$L^2(D)^3$	Abbreviation for $L^2(D, \mathbb{R})^3$, the space of square-integrable vector-valued (complex) functions on D. 8
$\|\mathbf{u}\|_2$	Norm of the space $L^2(D)^3$, $\|\mathbf{u}\|_2 := (\mathbf{u}, \mathbf{u})_D^{1/2}$. 17
$(\mathbf{u}, \mathbf{v})_D$	L^2-inner product on D, $(\mathbf{u}, \mathbf{v})_D := \int_D \mathbf{u} \cdot \mathbf{v}\,d\mathbf{x}$. 17
$W(D)$	Solution space for the magnetoquasistatic problem. 19
$W_{\text{st}}(D)$	Solution space for the magnetostatic problem. 19
$W_{2D}(D)$	Solution space for the two-dimensional magnetoquasistatic problem. 21
$\mathcal{H}_0(\mathbf{curl}, D)$	Space of functions $\mathbf{u} \in \mathcal{H}(\mathbf{curl}, D)$ subject to a zero Dirichlet boundary condition, i.e., $\mathbf{u} \times \mathbf{n} = 0$ on ∂D. 19
$\mathcal{H}_0(\mathbf{grad}, D)$	Space of functions $u \in \mathcal{H}(\mathbf{grad}, D)$ subject to a zero Dirichlet boundary condition, i.e., $u = 0$ on ∂D. 19

Chapter 1
Introduction

Nowadays, many physical phenomena and technical devices can be simulated by computers. Electromagnetic fields are fundamental for the understanding of electrotechnical components, such as antennas, electrical machines, transformers, electronic equipment and particle accelerators. Mathematical modeling is commonly based on partial differential equations, which are numerically approximated, e.g., by finite element, finite difference or finite volume methods. Modern numerical schemes can be designed with high accuracy and reliability in view of advances in, e.g., higher order modeling, error estimation and adaptivity and an ever-increasing computation power. Hence, the simulation of more complex phenomena and engineering designs is addressed. In this respect, the input data available becomes a major source of imprecision in simulations. Due to the manufacturing process and measurement errors, input data of the simulation, such as the geometry or material coefficients, is not known precisely. Uncertainty quantification and sensitivity analysis aim at further increasing the reliability of simulation predictions, by taking into account manufacturing imperfections or related variability associated to any real-life device.

A schematic view of the numerical simulation with uncertainties is depicted in Fig. 1.1. Uncertainties, e.g., in the geometry, arise at the input level of the model, a set of partial differential equations. The model is reformulated in a stochastic setting to account for input variations. Then, by solving the stochastic equations, uncertainties in the model output, a given field distribution or a physical quantity of interest of the system, are quantified.

Simulations with uncertain input data are more time consuming as each variable input parameter gives rise to an additional dimension in the associated parametric/stochastic model. In particular, possibly coupled systems of increased size must be solved. Uncertainty quantification addresses both an appropriate and low-dimensional modeling of the input uncertainties and their efficient propagation to the model outputs. Considerable effort has been spent in recent years concerning the design of spectral stochastic methods for uncertainty propagation. These schemes

U. Römer, *Numerical Approximation of the Magnetoquasistatic Model with Uncertainties*, Springer Theses, DOI 10.1007/978-3-319-41294-8_1

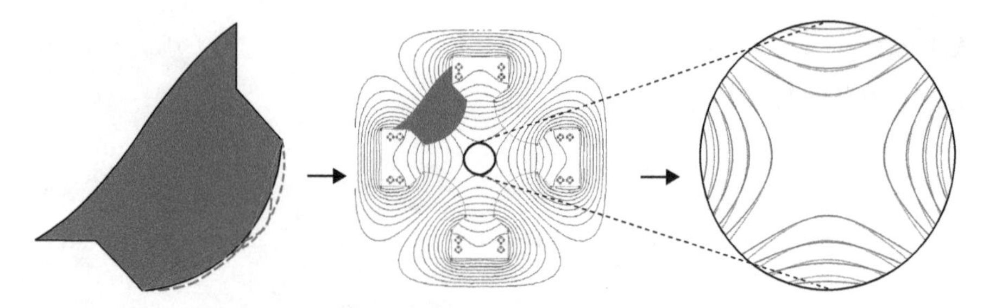

Fig. 1.1 A schematic view of uncertainty quantification in the numerical simulation (fields obtained using FEMM [1]) of a quadrupole magnet. Uncertainty modeling (*left-middle*): parametric/stochastic shape of a magnet pole as a simulation input. Uncertainty propagation (*middle-right*): parametric/stochastic simulations to quantify impact on model output, a Fourier-harmonic of the magnetic field. Deviations (*red*) from ideal field distribution (*grey*)

have been proven to outperform classical Monte Carlo techniques in many situations. In the case of small input deviations perturbation techniques are another cheap alternative.

Magnet design for particle accelerators is a challenging task, with high accuracy requirements. Uncertainties in the material properties (the $B - H$ curve), the shape of the magnet poles and the coils may affect the field homogeneity, quantified by means of Fourier-harmonics or multipole coefficients. Multipole fluctuations may result in undesired deviations of the particle beam from its designed trajectory. Quantifying uncertainties in this regard, is a promising tool in order to design magnets with increased reliability. Moreover, a precise estimation of tolerances might result in reduced production cost. As the magnetic energy is dominant for this type of applications, the electromagnetic fields are accurately described by means of the parabolic-elliptic magnetoquasistatic model. However, a major challenge with regard to this model is the nonlinearity in the magnetic constitutive law. Methods for uncertainty quantification and propagation require appropriate modifications in this respect.

The aim of the present work is the formulation and application of uncertainty quantification techniques for the nonlinear magnetoquasistatic formulation with application to magnet design. Its main achievements are: detailed sensitivity analysis results, in particular with respect to the material input data and shapes; the formulation of a moment based perturbation method published in [2]; a flexible and efficient modeling of stochastic $B - H$ curves in terms of the Karhunen-Loève expansion [3]; regularity results for a related parametric/stochastic model by means of a higher order sensitivity analysis [3]; the application of uncertainty quantification tools to magnet design, partially published in [4].

The thesis is structured as follows: in Chap. 2 the magnetoquasistatic approximation of Maxwell's equations is recalled as well as basics of uncertainty quantification. Numerical schemes for the approximation of the deterministic magnetoquasistatic model are addressed in Chap. 3. The model is adequately parametrized in Chap. 4, e.g., by means of splines for the $B - H$ curve and the geometry. This chapter also contains a detailed discussion of direct and adjoint sensitivity analysis techniques. In Chap. 5 a stochastic/parametric model will be derived and analyzed. In this respect probabilistic and non-probabilistic modeling will be addressed as well as the reduction of both input parameters and model dimensions. Based on sensitivity analysis tools, a moment based perturbation method will be derived and compared to the Monte Carlo method. A regularity analysis for the parametric/stochastic problem is carried out and a convergence result for the stochastic collocation method is established. Finally, Chap. 6 is devoted to uncertainty quantification in the context of magnet design with emphasis on numerical examples for dipole and quadrupole magnets. The thesis closes with conclusions and an outlook in Chap. 7.

References

1. Meeker, D.: Finite element method magnetics. Version 4.2 (1 April 2009 Build) (2010)
2. Römer, U., Schöps, S., Weiland, T.: Approximation of moments for the nonlinear magnetoquasistatic problem with material uncertainties. IEEE Trans. Magn. **50**(2) (2014)
3. Römer, U., Schöps, S., Weiland, T.: Stochastic modeling and regularity of the nonlinear elliptic curl-curl equation. SIAM/ASA J Uncertainty Quantification (in press)
4. Bartel, A., De Gersem, H., Hülsmann, T., Römer, U., Schöps, S., Weiland, T.: Quantification of uncertainty in the field quality of magnets originating from material measurements. IEEE Trans. Magn. **49**(5), 2367–2370 (2013)

Chapter 2
Magnetoquasistatic Approximation of Maxwell's Equations, Uncertainty Quantification Principles

Starting from the classical form of Maxwell's equations the magnetoquasistatic approximation will be derived and justified. Additionally, some key notions from the area of uncertainty quantification, verification and validation will be established.

2.1 Maxwell's Equations

Here, we adopt the classical 3-D Euclidean vector representation, as opposed to the 4-D space-time form of electromagnetics based on exterior calculus and differential forms. The content of this section can be found in many textbooks, see, e.g., [1, 2]. Maxwell's equations in integral form read as

$$\int_{\partial V} \mathbf{D} \cdot \mathrm{d}\mathbf{A} = \int_V \rho \mathrm{d}V, \tag{2.1a}$$

$$\int_{\partial A} \mathbf{H} \cdot \mathrm{d}\mathbf{s} = \int_A \left(\mathbf{J} + \frac{\partial \mathbf{D}}{\partial t} \right) \cdot \mathrm{d}\mathbf{A}, \tag{2.1b}$$

$$\int_{\partial V} \mathbf{B} \cdot \mathrm{d}\mathbf{A} = 0, \tag{2.1c}$$

$$\int_{\partial A} \mathbf{E} \cdot \mathrm{d}\mathbf{s} = -\int_A \frac{\partial \mathbf{B}}{\partial t} \cdot \mathrm{d}\mathbf{A}, \tag{2.1d}$$

for any surface $A \subset \mathbb{R}^3$ and volume $V \subset \mathbb{R}^3$ at rest. These relations contain the electric induction $\mathbf{D}$, the electric charge density ρ, the magnetic field $\mathbf{H}$, the magnetic induction $\mathbf{B}$, the electric field $\mathbf{E}$ and the electric current density $\mathbf{J}$. The current density is decomposed as $\mathbf{J} = \mathbf{J}_{\mathrm{src}} + \mathbf{J}_{\mathrm{con}}$, where $\mathbf{J}_{\mathrm{src}}$ and $\mathbf{J}_{\mathrm{con}}$ refer to an imposed and ohmic part, respectively. Relations (2.1) have to be supplemented by material constitutive relations. For time-invariant, isotropic media these are given by

U. Römer, *Numerical Approximation of the Magnetoquasistatic Model with Uncertainties*, Springer Theses, DOI 10.1007/978-3-319-41294-8_2

$$\mathbf{D} = \varepsilon_0 \mathbf{E} + \mathbf{P}, \tag{2.2a}$$

$$\mathbf{B} = \mu_0 (\mathbf{H} + \mathbf{M}), \tag{2.2b}$$

$$\mathbf{J}_{\text{con}} = \sigma \mathbf{E}, \tag{2.2c}$$

where ε_0, μ_0 represent the permittivity and permeability of vacuum, σ refers to the electric conductivity and **P**, **M** represent the electric polarization and the magnetization, respectively. Except for ε_0, μ_0, all quantities in (2.2) are functions of space. Moreover, **M** and **P** depend on **H** and **E**, respectively. Using the theorems of Gauss and Stokes we derive from (2.1) the differential form of Maxwell's equations

$$\operatorname{div} \mathbf{D} = \rho, \tag{2.3a}$$

$$\mathbf{curl}\ \mathbf{H} = \mathbf{J} + \frac{\partial \mathbf{D}}{\partial t}, \tag{2.3b}$$

$$\operatorname{div} \mathbf{B} = 0, \tag{2.3c}$$

$$\mathbf{curl}\ \mathbf{E} = -\frac{\partial \mathbf{B}}{\partial t}, \tag{2.3d}$$

endowed with suitable boundary conditions or decay conditions at infinity if the domain is bounded or unbounded, respectively. The integral version of Maxwell's equations is often the starting point instead of (2.3) as they are formulated by means of directly measurable quantities such as voltages and fluxes. The key observation here, is that $\int_S \mathbf{E} \cdot \mathrm{d}\mathbf{s}$ can actually be viewed as a mapping

$$S \mapsto \int_S \mathbf{E} \cdot \mathrm{d}\mathbf{s} \tag{2.4}$$

which associates a voltage to a sufficiently smooth oriented line. This is precisely the notion of an integral form of degree one, equivalently

$$A \mapsto \int_A \mathbf{B} \cdot \mathrm{d}\mathbf{A} \tag{2.5}$$

is an integral form of degree two. The notion of integral forms will be used later on, to derive boundary and interface conditions.

2.2 Magnetoquasistatic Approximation

Although practically all phenomena of classical electromagnetics are governed by Maxwell's equations, their resolution, in particular their numerical resolution, in the most general form is often unnecessary, sometimes impracticable. In this work we are especially interested in simulating devices with dominant magnetic energy, where wave propagating effects can be neglected. As we will justify, this amounts

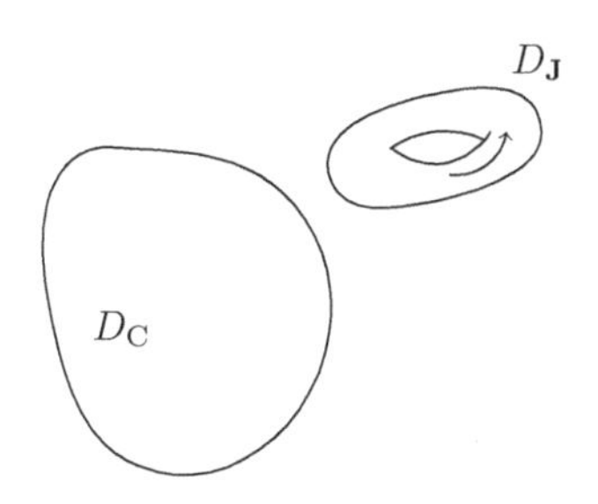

Fig. 2.1 Model geometry for the illustration of the magnetoquasistatic approximation on an unbounded domain

in neglecting the term $\partial\mathbf{D}/\partial t$ in (2.3). Typical applications are electrical machines and transformers operating at low frequencies and accelerator magnets, which are the key application in this work. The resulting set of equations is referred to as *magnetoquasistatic* model or eddy current model in the literature.

To simplify the discussion we restrict ourselves to the time-harmonic equations in the remaining part of this subsection. A model geometry is depicted in Fig. 2.1. We consider a bounded, connected domain of homogeneous conductivity D_C in free space and an imposed divergence free current density with $\mathrm{supp}(\mathbf{J}_{\mathrm{src}}) =: D_J$. We assume that $\overline{D_J} \cap \overline{D_C} = \emptyset$ and that σ is constant inside D_C and zero in free space $D_E = \mathbb{R}^3 \backslash \overline{D_C}$, whereas μ and ε are assumed to be piecewise constant. The restrictions of the material functions to D_C, D_E are denoted with subscripts C, E, respectively. Following [3], the time-harmonic magnetoquasistatic model is given by

$$\mathrm{div}\,\varepsilon\mathbf{E} = 0, \qquad \text{in } D_E, \tag{2.6a}$$

$$\mathbf{curl}\ \mathbf{H} = \sigma\mathbf{E} + \mathbf{J}_{\mathrm{src}}, \qquad \text{in } \mathbb{R}^3, \tag{2.6b}$$

$$\mathbf{curl}\ \mathbf{E} = -j\omega\mu\mathbf{H}, \qquad \text{in } \mathbb{R}^3, \tag{2.6c}$$

$$\mathbf{E}(\mathbf{x}) = \mathcal{O}\left(\frac{1}{|\mathbf{x}|^2}\right), \qquad \text{uniformly for } |\mathbf{x}| \to \infty, \tag{2.6d}$$

$$\mathbf{H}(\mathbf{x}) = \mathcal{O}\left(\frac{1}{|\mathbf{x}|^2}\right), \qquad \text{uniformly for } |\mathbf{x}| \to \infty, \tag{2.6e}$$

where j is the imaginary unit and $|\cdot|$ refers to the Euclidean norm. Uniqueness is achieved, by additionally imposing the condition $\int_{\partial D_C} \mathbf{E}\cdot \mathrm{d}\mathbf{A} = 0$. Often, neglecting the displacement current $j\omega\mathbf{D}$ is justified as a low frequency approximation, i.e., $\omega \to 0$. In this case the modeling error, i.e., the difference between the electromagnetic fields described by Maxwell's equations ($\mathbf{E}^{\mathrm{M}}$, $\mathbf{H}^{\mathrm{M}}$) (M refers to full Maxwell) and the magnetoquasistatic equations ($\mathbf{E}$, $\mathbf{H}$), has been analyzed in [3]. In a first step it can be shown, based on a power series expansion $\mathbf{E}^{\mathrm{M}} = \omega\mathbf{E}_1 + \mathcal{O}(\omega^2)$, that

$$\mathbf{curl}\ \mathbf{H}^{\mathrm{M}} = \sigma\omega\mathbf{E}_1 + \mathbf{J}_{\mathrm{src}} + \mathcal{O}(\omega^2), \tag{2.7}$$

see [3], thus the neglected term is of higher order in ω. Moreover,

$$\|\mathbf{E}^{\mathrm{M}} - \mathbf{E}\|_{L^2(B_R)^3} = \mathcal{O}(\omega^2), \quad \|\mathbf{H}^{\mathrm{M}} - \mathbf{H}\|_{L^2(B_R)^3} = \mathcal{O}(\omega^2), \tag{2.8}$$

holds, see [3], where $D_{\mathrm{J}} \subset B_R \subset \mathbb{R}^3$ and B_R is a ball of radius R, provided that $\mathbf{J}_{\mathrm{src}}$ is divergence free in the limit. The asymptotic behavior (2.8) has been confirmed for a bounded domain in [4]. However, if D_{J} and D_{C} overlap, i.e., there exists a galvanic connection expressed as a current flow to the conductor, the asymptotic error decay reduces to $\mathcal{O}(\omega)$.

Although appropriate in many cases the notion of low-frequency approximation is not general enough, as it is well known that in several circumstances the magnetoquasistatic model can be used for moderate up to high frequencies, too. Let us therefore consider the more general, and widely used condition, that

$$\omega\varepsilon_{\mathrm{C}} \ll \sigma_{\mathrm{C}} \tag{2.9}$$

has to be small. For a bounded domain D, we supplement this condition by

$$\mathrm{diam}(D) \ll \lambda, \tag{2.10}$$

where diam refers to the diameter of D. This means that the dimensions of the domain D have to be small compared to the wavelength $\lambda = 2\pi/(\sqrt{\mu\epsilon}\omega)$. Again, these conditions can be mathematically justified, as done in [4], where the relation

$$\frac{\|\mathbf{E}^{\mathrm{M}} - \mathbf{E}\|_{L^2(D)^3}}{\|\mathbf{E}^{\mathrm{M}}\|_{L^2(D)^3}} \leq C_1 \frac{\mathrm{diam}(D)^2}{\lambda_{\mathrm{C}}^2} + C_2 \frac{\omega\varepsilon_{\mathrm{C}}}{\sigma_{\mathrm{C}}} \tag{2.11}$$

was derived, with positive constants $C_1, C_2 > 0$. However, one should keep in mind, that even if (2.9) and (2.10) are satisfied the magnetoquasistatic approximation might still be unjustified as the constants C_1, C_2 in the previous expressions can become quite large in some situations: the conductor geometry has to be such that capacitive effects are negligible [4, 5].

2.3 Magnetoquasistatic Model

Although the magnetoquasistatic model was justified in a time-harmonic setting in the previous section, we will choose a time-domain setting from now on, due to the following reasons. The magnetic material law is typically nonlinear and higher order harmonics of the fields need to be taken into account, even if the current is excited at a single frequency. This can be done, and actually is a popular choice for the simulations of electrical machines [6]. It should also be noted that usually a few higher order harmonics are sufficient and the truncation error is well understood [7]. However, as a more serious drawback, a time-harmonic setting is inappropriate for modeling strongly time transient phenomena we want to include into our setting, such as the ramping of a magnet. We are now going to introduce in some detail the model problem that, under simplifications from time to time, will be the basis for the remaining part of this work. For many applications it is acceptable to consider a bounded computational domain D. This reflects the fact, that the fields decay as given in (2.6d), (2.6e) and the energy stored in a region far away from the current excitation is close to zero. In this context, the shape of D has no physical significance and we assume that D is a simply connected polyhedral Lipschitz domain. Lipschitz boundaries are needed to apply many results on Sobolev spaces, whereas polyhedral domains, i.e., domains with boundaries consisting of plain faces, straight edges and corner points, can be exactly covered by conventional tetrahedral finite element meshes. For a precise definition of these terms, see, e.g., [8, 9]. The model geometry is depicted in Fig. 2.2, where D (strictly) contains a conducting, ferromagnetic region D_C and an air region filled with coil parts, defined as D_E such that $D_\mathrm{E} = D\backslash\overline{D_\mathrm{C}}$. We do not assume that D_C is simply connected, as this would exclude many applications. Concerning the constitutive relations the effects of isotropy and hysteresis and permanent magnetization are neglected here. Although this might oversimplify many practical setups, it allows for a more thorough modeling of uncertainties, which is a main issue of this work. More precisely, we assume the following:

Assumption 2.1 (*Conductivity*) The electric conductivity satisfies

$$\sigma(\mathbf{x}) = \begin{cases} \sigma_\mathrm{C}, & \text{in } D_\mathrm{C}, \\ 0, & \text{in } D_\mathrm{E}, \end{cases} \tag{2.12}$$

where $\sigma_\mathrm{C} > 0$ is supposed to be constant.

For nonlinear materials, the magnetic properties at each point are expressed through the so called $B - H$ curve $f_{BH} : \mathbb{R}_0^+ \to \mathbb{R}_0^+$, given by

$$|\mathbf{B}| = f_{BH}(|\mathbf{H}|) := \mu_0(|\mathbf{H}| + |\mathbf{M}|). \tag{2.13}$$

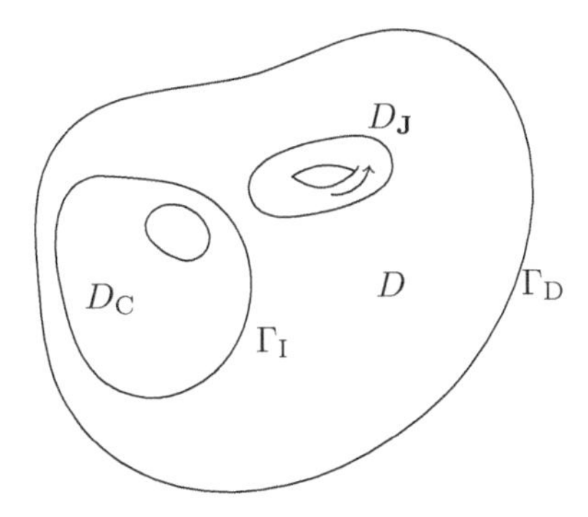

Fig. 2.2 General model geometry for the magnetoquasistatic model on a bounded domain

Let $C^1(\mathbb{R}^+)$ denote the space of continuously differentiable functions on $\mathbb{R}^+$. Well-known physical properties of $f_{BH} \in C^1(\mathbb{R}^+)$ are expressed as

$$f_{BH}(0) = 0, \tag{2.14a}$$
$$\partial_s f_{BH}(s) \geq \mu_0, \quad \forall s \geq 0, \tag{2.14b}$$
$$\lim_{s\to\infty} \partial_s f_{BH}(s) = \mu_0, \tag{2.14c}$$

see [10, 11]. From (2.14) it can be deduced that f_{BH} is a bijective function and the inverse $f_{HB} := f_{BH}^{-1}$ satisfies properties similar to (2.14). This motivates the following assumption for the magnetic reluctivity, defined point-wise as $\nu(s) := f_{HB}(s)/s,\ \forall s \in \mathbb{R}^+$ ($\nu(0)$ is defined by taking the limit $s \to 0$):

Assumption 2.2 (*Reluctivity*) The magnetic reluctivity $\nu : D \times \mathbb{R}_0^+ \to \mathbb{R}$ satisfies

$$\nu(\mathbf{x}, \cdot) = \begin{cases} \nu_{\mathrm{C}}(\cdot), & \text{in } D_{\mathrm{C}}, \\ \nu_0, & \text{in } D_{\mathrm{E}}, \end{cases} \tag{2.15}$$

with $\mathbf{x} \in D$ and value $\nu_0 = 1/\mu_0$, the reluctivity of vacuum. Moreover, $\nu_{\mathrm{C}} : \mathbb{R}_0^+ \to \mathbb{R}^+$ is a continuous function such that for all $s \in \mathbb{R}_0^+$,

$$0 < \nu_{\min} \leq \nu_{\mathrm{C}}(s) \leq \nu_0 < \infty, \tag{2.16a}$$
$$\nu_{\mathrm{C}}(\cdot)\cdot, \text{ is Lipschitz continuous with constant } \nu_0, \tag{2.16b}$$
$$\nu_{\mathrm{C}}(\cdot)\cdot, \text{ is strongly monotone with constant } \nu_{\min}, \tag{2.16c}$$

holds.

More precisely, Lipschitz continuity and strong monotonicity for $\nu_{\mathrm{C}}(\cdot)\cdot$ or f_{HB} are expressed as

$$|f_{HB}(s) - f_{HB}(t)| \leq \nu_0 |s - t|, \tag{2.17}$$
$$(f_{HB}(s) - f_{HB}(t))\,(s - t) \geq \nu_{\min}(s - t)^2, \tag{2.18}$$

for all $s, t \in \mathbb{R}^+$, respectively. In a more general setting equations (2.16) would hold for each $\mathbf{x} \in D$ and additionally $\nu(\cdot, s)$ would be assumed to be measurable for all $s \in \mathbb{R}_0^+$, see [12]. The assumption $\nu = \nu_{\mathrm{E}}$ in D_{E} is justified as the magnetic properties of coil parts, such as copper, can be well approximated by the respective vacuum properties. Working with the magnetic reluctivity and the inverse function f_{HB}, is particularly appropriate for the magnetic vector potential formulation. There exists a vast literature on the magnetoquasistatic model, see, e.g., [7, 13, 14]. Among the different formulations, the vector potential formulation is a rather general approach, well suited for the applications covered in this treatise and derived formally as follows.

As $\mathbf{B}$ is divergence free (2.1c), it can be represented by a vector potential $\mathbf{B} = \mathbf{curl}\ \mathbf{A}$. Then from (2.1b) and the magnetoquasistatic approximation we obtain

$$\sigma \mathbf{E} + \mathbf{curl}\ (\nu\, \mathbf{curl}\ \mathbf{A}) = \mathbf{J}_{\mathrm{src}}. \tag{2.19}$$

From Eq. (2.1d) in turn we infer $\partial \mathbf{A}/\partial t = \mathbf{E}$ (up to a gradient field) and hence the differential equation

$$\sigma \frac{\partial \mathbf{A}}{\partial t} + \mathbf{curl}\ (\nu\, \mathbf{curl}\ \mathbf{A}) = \mathbf{J}_{\mathrm{src}}, \tag{2.20}$$

which is valid, separately in each subdomain. To achieve uniqueness in the vector potential representation a gauging procedure is required. In particular, any gradient field can be added to $\mathbf{A}$ without changing $\mathbf{B}$, as $\mathbf{curl}\ \mathbf{grad}\ = 0$. To this end

$$\operatorname{div} \mathbf{A} = 0, \text{ in } D_{\mathrm{E}}, \quad \int_{\Gamma_{\mathrm{E},i}} \mathbf{A} \cdot \mathbf{n}\ \mathrm{d}\mathbf{x} = 0, \tag{2.21}$$

with exterior unit normal $\mathbf{n}$, is imposed for all connected components $\Gamma_{\mathrm{E},i} \subset \partial D_{\mathrm{E}}$. Equation (2.21) is an extension of the Coulomb gauge to multiply connected domains, see [15, 16] and also [17] for a rigorous treatment of vector potentials in a more general context. In D_{C}, $\mathbf{A}$ will automatically satisfy these conditions. This can be seen by taking the divergence of (2.19), provided that additionally both the current excitation and the initial condition for the vector potential are divergence free. To close the setting, the interface conditions at Γ_{I} have to be specified. The patch condition [18] implies that, in order for $\mathbf{A}$ and $\mathbf{H}$ to give rise to a valid integral 1-form, for all $S \subset D$,

$$\mathbf{n} \times (\mathbf{A}^{+} - \mathbf{A}^{-}) = 0, \tag{2.22a}$$
$$\mathbf{n} \times (\mathbf{H}^{+} - \mathbf{H}^{-}) = \mathbf{n} \times ((\nu\, \mathbf{curl}\ \mathbf{A})^{+} - (\nu\, \mathbf{curl}\ \mathbf{A})^{-}) = 0. \tag{2.22b}$$

Here, we denote with $\mathbf{H}^{+}$ and $\mathbf{H}^{-}$, the restriction of $\mathbf{H}$ to S, from the exterior and interior (with respect to $\mathbf{n}$), respectively. In this setting we exclude the presence of surface currents, that can be incorporated in a more general, distributional setting [19]. As interface conditions, such as (2.22), play an important role in this work we also introduce the operator

$$(\mathbf{A})_S := \mathbf{n} \times (\mathbf{A}^{+} - \mathbf{A}^{-}). \tag{2.23}$$

Let $I_T = (0, T]$ denote the time interval of interest. In summary, we want to determine the magnetic vector potential $\mathbf{A}(t, \mathbf{x})$ subject to

$$\sigma(\cdot) \frac{\partial \mathbf{A}}{\partial t} + \mathbf{curl}\ (\nu(\cdot, |\mathbf{curl}\ \mathbf{A}|)\, \mathbf{curl}\ \mathbf{A}) = \mathbf{J}_{\mathrm{src}}, \qquad \text{in } I_T \times (D_{\mathrm{C}} \cup D_{\mathrm{E}}), \tag{2.24a}$$

$$\mathbf{n} \times (\mathbf{A}_\mathrm{C} - \mathbf{A}_\mathrm{E}) = 0, \qquad \text{on } I_T \times \Gamma_\mathrm{I}, \tag{2.24b}$$

$$\mathbf{n} \times (\nu_\mathrm{C}(|\,\mathbf{curl}\,\mathbf{A}_\mathrm{C}|)\,\mathbf{curl}\,\mathbf{A}_\mathrm{C} - \nu_0\,\mathbf{curl}\,\mathbf{A}_\mathrm{E}) = 0, \qquad \text{on } I_T \times \Gamma_\mathrm{I}, \tag{2.24c}$$

$$\mathbf{A} \times \mathbf{n} = 0, \qquad \text{on } I_T \times \Gamma_\mathrm{D}, \tag{2.24d}$$

$$\mathbf{A}(0) = \mathbf{A}_\mathrm{init}, \qquad \text{on } \{0\} \times D, \tag{2.24e}$$

$$\mathrm{div}\,\mathbf{A} = 0, \qquad \text{in } I_T \times D_\mathrm{E}, \tag{2.24f}$$

$$\int_{\Gamma_{\mathrm{E},i}} \mathbf{A} \cdot \mathbf{n}\ \mathrm{d}\mathbf{x} = 0, \qquad \text{on } I_T, \tag{2.24g}$$

where $\mathbf{A}_E = \mathbf{A}|_{D_\mathrm{E}}$ and $\mathbf{A}_C = \mathbf{A}|_{D_\mathrm{C}}$ and Γ_I is oriented such that $\mathbf{n}$ is the exterior unit normal with respect to the domain D_C. To simplify notation the spatial dependency is omitted when possible, i.e., $\mathbf{A}(t) := \mathbf{A}(t, \mathbf{x})$. We observe that equation (2.24) is an initial boundary value problem of parabolic-elliptic type, parabolic in D_C and elliptic in D_E. Let us note that gauging is rather simple in our case, due the fact that σ_C is constant. For the general case of varying σ_C we refer, e.g., to [20] and for a more detailed discussion to [21]. Also, a similar formulation holds true for the electric field. From now on, for simplicity, the electric current density $\mathbf{J}$ always refers to the imposed part, i.e., we set $\mathbf{J} = \mathbf{J}_\mathrm{src}$.

2.4 Uncertainty Quantification, Verification and Validation

The aim of this section is to give a brief introduction to uncertainty quantification as well as verification and validation and to specify the required terminology. The exposition will be based mainly on work from mechanical engineering [22–25] and the monograph [26], that we consider to be the most appropriate for our purposes.

- Following [22], a (mathematical) *model* is defined as a "collection of mathematical constructions that provide abstractions of a physical event consistent with a scientific theory proposed to cover that event". In the context of this work, this refers to the magnetoquasistatic model, i.e., the system of nonlinear partial differential equations given by (2.24). Model input data is identified with "data from the description of the surroundings", see [23]. In general this refers to initial and boundary conditions, shapes and material constitutive laws. A *model output* in the general case is given by the solution of the underlying system of differential equations. However, in many practical situation not the field itself is used within the design process but some derived *quantities of interest*. Mathematically, these are given as functionals, i.e., maps from the solution to the real numbers. Important examples, are the magnetic energy, the inductance, power losses and most notably in our case multipole coefficients, i.e., Fourier coefficients of the magnetic field.
- Despite the fact that the magnetoquasistatic model may provide a very accurate mathematical description of reality, it is always a simplification of the real underlying physics. Assessing this discrepancy is commonly denoted as *validation* [22, 24, 26]. This comprises both the comparison with measurements and the quantitative

estimation of modeling errors by means of a posteriori error estimation, as outlined, e.g., in [27]. We also refer to [28] for the important case of a posteriori error estimation of the linearization error. Also the arguments given in Sect. 2.2 to justify the magnetoquasistatic approximation can be assigned to the process of validation.

- A set of partial differential equations posed on complicated domains, cannot be resolved directly in general and approximations have to be introduced. By means of numerical approximations, a *computational model*, i.e., a linear system of equations solved by a computer, is derived [22]. Several different types of approximation errors occur at this stage. These are discretization errors, round-off errors as well as errors from the numerical resolution of the linear system of equations. Considerable progress has been made in controlling most of them by means of a posteriori error analysis. We refer to [29] and the references therein for an overview of discretization error estimation in a finite element context. This error contribution, as well as the linearization error, will be of central importance in this work. Note that several errors such as coding errors, might not even be known. The general process of evaluating whether an implemented computational model can be used to accurately represent the mathematical model is referred to as *verification*.
- The setting introduced so far is completely deterministic, in the sense that the input data is considered to be known exactly and to each input is associated a solution by the computational model. This view has several shortcomings with respect to depicting real life devices and machines. Indeed, in practice uncertainties arise and should be incorporated in several different parts of the model. Every single part of a device, as produced from chain production, has a different material composition and shape. Furthermore, it is often unknown whether the chosen form of the model is appropriate to describe the underlying physics. Consequently, model inputs as well as the model form exhibit uncertainties and any reliable design demands for their quantification. In the literature two types of uncertainties are widely acknowledged: *aleatory* and *epistemic* uncertainty [23]. Aleatory uncertainty is defined as "the inherent variation associated with the physical system or the environment under consideration", see [26]. It originates usually from manufacturing imperfections and is considered irreducible for the system under consideration [23]. Epistemic uncertainty is defined as "any lack of knowledge of information in any phase or activity of the modeling process", [26]. In contrast to aleatory uncertainty, epistemic uncertainty might be considered reducible as, e.g., by measurements the belief in a specific model might be increased. Mathematically, aleatory uncertainty is described in the most general form by a probability density function, whereas non-probabilistic quantities subject to epistemic uncertainty, will belong to an admissible set with equal probability of occurrence for each element. Note that in practice distinguishing between both types may be difficult, sometimes rather subjective.
- If a mathematical description of the input uncertainties is at hand, another important step of uncertainty quantification consists in propagating them through the model. Thereby, even more errors, consisting of both modeling and numerical errors occur

as, e.g., solving stochastic equations might quickly become very costly and simplifications are required. As soon as the output uncertainties are quantified, strong sensitivities may be taken into account to increase the robustness of the design.

Remark 2.1 There is an ambiguity in the literature, whether numerical error should be considered as epistemic uncertainty. In [25, 26], numerical error is considered to be a "recognizable deficiency in any phase or activity of modeling that is not due to the lack of knowledge", whereas in [23] it is argued that for complex systems this might be not practicable and error cannot always be identified and reduced. We adapt the latter view, as it might be useful to compare uncertainties and errors in the context of error balancing.

Another important aspect is that uncertainty quantification is not feasible for any kind of mathematical model. Following [26] we identify as minimal requirements for the mathematical model, the existence of a unique solution as well as a continuous dependence of the solution, or output, on the input data. Models that feature these properties are denoted *well-posed* after Hadamard. Indeed, especially the continuity can be seen as essential for the purpose of uncertainty quantification, as it assures that small changes in the input produce small changes in the output. Further, highly desirable, model features are the existence of a numerical solution as well as the differentiability of the model outputs with respect to the model inputs [26]. In the latter case, sensitivity analysis can be used to efficiently propagate uncertainties.

2.5 Conclusion

So far, the magnetoquasistatic approximation to Maxwell's equations was derived and justified. Key assumptions on data, such as reluctivity, conductivity and geometry were specified. Finally, notions from verification and validation, in particular concerning uncertainty quantification, were introduced and discussed.

References

1. Hehl, F.W., Obukhov, I.N.: *Foundations of Classical Electrodynamics: Charge, Flux, and Metric*, vol. 33. Springer (2003)
2. Jackson, J.D.: *Classical Electrodynamics*, 3rd edn. Wiley, New York (1999)
3. Buffa, A., Ammari, H., Nédélec, J.-C.: A justification of eddy currents model for the Maxwell equations. SIAM J. Appl. Math. **60**(5), 1805–1823 (2000)
4. Schmidt, K., Sterz, O., Hiptmair, R.: Estimating the eddy-current modeling error. IEEE Trans. Magn. **44**(6), 686–689 (2008)
5. Bossavit, A.: Computational Electromagnetism: Variational Formulations, Complementarity, Edge Elements. Academic Press, San Diego (1998)
6. Gyselinck, J., Dular, P., Geuzaine, C., Legros, W.: Harmonic-balance finite-element modeling of electromagnetic devices: a novel approach. IEEE Trans. Magn. **38**(2), 521–524 (2002)

7. Bachinger, F., Langer, U., Schöberl, J.: Numerical analysis of nonlinear multiharmonic eddy current problems. Numerische Mathematik **100**(4), 593–616 (2005)
8. Monk, P.: *Finite Element Methods for Maxwell's Equations*. Oxford University Press (2003)
9. Delfour, M.C., Zolésio, J.-P.: *Shapes and Geometries: Metrics, Analysis, Differential Calculus, and Optimization*, 1 edn. SIAM (2001)
10. Reitzinger, S., Kaltenbacher, B., Kaltenbacher, M.: A note on the approximation of B-H curves for nonlinear computations. Technical Report 02-30, SFB F013, Johannes Kepler University Linz, Austria (2002)
11. Pechstein, C.: Multigrid-newton-methods for nonlinear magnetostatic problems. M.Sc. thesis, Johannes Kepler Universität Linz, Austria (2004)
12. Yousept, I.: Optimal control of quasilinear $H(curl)$-elliptic partial differential equations in magnetostatic field problems. SIAM J. Control Optim. **51**(5), 3624–3651 (2013)
13. Carpenter, C.J.: Comparison of alternative formulations of 3-dimensional magnetic-field and eddy-current problems at power frequencies. Proc. Inst. Electr. Eng. **124**(11), 1026–1034 (1977)
14. Rodríguez, A.A., Valli, A.: *Eddy Current Approximation of Maxwell Equations: Theory, Algorithms and Applications*, vol. 4. Springer (20100
15. Fernandes, P., Perugia, I.: Vector potential formulation for magnetostatics and modelling of permanent magnets. IMA J. Appl. Math. **66**(3), 293–318 (2001)
16. Bossavit, A.: Magnetostatic problems in multiply connected regions: some properties of the curl operator. IEE Proc. A **135**(3), 179–187 (1988)
17. Amrouche, C., Bernardi, C., Dauge, M., Girault, V.: Vector potentials in three-dimensional non-smooth domains. Math. Methods Appl. Sci. **21**(9), 823–864 (1998)
18. Hiptmair, R.: Finite elements in computational electromagnetism. Acta Numerica **11**, 237–339 (2002)
19. Cessenat, M.: *Mathematical Methods In Electromagnetism*. World Scientific (1996)
20. Arnold, L., von Harrach, B.: A unified variational formulation for the parabolic-elliptic eddy current equations. SIAM J. Appl. Math. **72**(2), 558–576 (2012)
21. Kettunen, L., Forsman, K., Bossavit, A.: Gauging in whitney spaces. IEEE Trans. Magn. **35**(3), 1466–1469 (1999)
22. Babuška, I., Oden, J.T.: Verification and validation in computational engineering and science: basic concepts. Comput. Methods Appl. Mech. Eng. **193**(36), 4057–4066 (2004)
23. Roy, C.J., Oberkampf, W.L.: A comprehensive framework for verification, validation, and uncertainty quantification in scientific computing. Comput. Methods Appl. Mech. Eng. **200**(25), 2131–2144 (2011)
24. Schwer, L.E.: Guide for verification and validation in computational solid mechanics. Am. Soc. Mech. Eng. (2006)
25. Oberkampf, W.L., Helton, J.C., Sentz, K.: Mathematical representation of uncertainty. In: *AIAA Non-Deterministic Approaches, Forum*, pp. 16–19 (2001)
26. Hlaváček, I., Chleboun, J., Babuška, I.: *Uncertain Input Data Problems and the Worst Scenario Method*. Elsevier (2004)
27. Oden, T.J., Prudhomme, S.: Estimation of modeling error in computational mechanics. J. Comput. Phys. **182**(2), 496–515 (2002)
28. Chaillou, A.L., Suri, M.: Computable error estimators for the approximation of nonlinear problems by linearized models. Comput. Methods Appl. Mech. Eng. **196**(1), 210–224 (2006)
29. Ainsworth, M., Oden, J.T.: A posteriori error estimation in finite element analysis. Comput. Methods Appl. Mech. Eng. **142**(1), 1–88 (1997)

Chapter 3
Magnetoquasistatic Model and its Numerical Approximation

We proceed with a more mathematical and detailed investigation of the magnetoquasistatic model as presented in Chap. 2. After establishing a weak formulation, we consider approximation in space, linearization of the nonlinear model and discretization with respect to time, respectively. Most of the content presented here is not new but needed for the remaining part of the work. An exception is the a posteriori error estimation of discretization and linearization error in Sect. 3.5 which extends work given for the nonlinear Poisson equation [1].

3.1 Weak Formulation

We are now going to derive a weak formulation in order to show that (2.24) possesses a unique solution and to introduce a finite element discretization. To this end we formally multiply (2.24a) with a test function $\mathbf{v}$ and integrate over D. Integration by parts yields

$$\int_D \sigma \frac{\partial \mathbf{A}}{\partial t} \cdot \mathbf{v}\, \mathrm{d}\mathbf{x} + \int_D \nu(\cdot, |\mathbf{curl}\, \mathbf{A}|)\mathbf{curl}\, \mathbf{A} \cdot \mathbf{curl}\, \mathbf{v}\, \mathrm{d}\mathbf{x} = \int_D \mathbf{J} \cdot \mathbf{v}\, \mathrm{d}\mathbf{x}. \tag{3.1}$$

A definition of the function spaces for $\mathbf{A}$ and $\mathbf{v}$ in view of (2.24) is in order. For $\mathbf{u}, \mathbf{v} \in L^2(D)^3$ we introduce the inner product

$$(\mathbf{u}, \mathbf{v})_D := \int_D \mathbf{u} \cdot \mathbf{v}\, \mathrm{d}\mathbf{x} \tag{3.2}$$

and the norm $\|\mathbf{u}\|_2 := (\mathbf{u}, \mathbf{u})_D^{1/2}$. Moreover, we define $\mathcal{H}(\mathbf{grad}, D)$ as the space of functions $u \in L^2(D)$ with (weak) gradient $\mathbf{grad}\, u \in L^2(D)^3$. Weak divergence free functions on multiply connected domains are defined, following [2], as:

U. Römer, *Numerical Approximation of the Magnetoquasistatic Model with Uncertainties*, Springer Theses, DOI 10.1007/978-3-319-41294-8_3

Definition 3.1 (*Weak Divergence Free Function*) Let V be a multiply connected domain, with r components of the boundary ∂V, denoted as $\Gamma_{V,i}$. We set

$$G(V) := \{\mathbf{v} = \mathbf{grad}\,\varphi \mid \varphi \in \mathcal{H}(\mathbf{grad}, V), \varphi = c_i \text{ on } \Gamma_{V,i}, 1 \leq i \leq r\}. \tag{3.3}$$

A function $\mathbf{v} \in L^2(V)^3$ satisfying

$$(\mathbf{v}, \mathbf{w})_V = 0, \ \forall \mathbf{w} \in G(V), \tag{3.4}$$

is called weak divergence free in V.

For a simply connected domain with $r = 1$ we set $c_1 = 0$. If $\mathbf{v}$ satisfies (3.4) it can be shown that $\operatorname{div} \mathbf{v} = 0$ almost everywhere in V and $\int_{\Gamma_{V,i}} \mathbf{v} \cdot \mathbf{n} = 0$ for all connected components $\Gamma_{V,i} \subset \partial V$. We recall that a condition is said to hold almost everywhere (a.e.) on a set V, if it holds after possible modification on a subset of measure zero, see [3, p. 15]. In a next step we formalize the assumptions on the source current and the initial condition.

Assumption 3.1 (*Divergence Free Data*) The electric source current density $\mathbf{J}(t) \in L^2(D_\mathrm{E})^3$, satisfies for all $t \in (0, T]$

$$\int_{D_\mathrm{E}} \mathbf{J}(t) \cdot \mathbf{v} \,\mathrm{d}\mathbf{x} = 0, \quad \forall t \in (0, T], \ \forall \mathbf{v} \in G(D_\mathrm{E}). \tag{3.5}$$

The initial condition $\mathbf{A}_\mathrm{init} \in L^2(D)$ satisfies

$$\int_{D_\mathrm{C}} \mathbf{A}_\mathrm{init} \cdot \mathbf{v} \,\mathrm{d}\mathbf{x} = 0, \quad \forall \mathbf{v} \in G(D_\mathrm{C}). \tag{3.6}$$

It remains to take into account the specific situation that the solution requires a gauging in D_E but not in D_C. This has been addressed in [2] by using a Schur complement approach. It relies on the observation, that in D_C the problem is uniquely solvable without gauging and that $\mathbf{A}_\mathrm{E}$ in turn is uniquely determined by $\mathbf{A}_\mathrm{C}$. Mathematically the latter relation is expressed through the following mapping [2]:

Definition 3.2 (*Harmonic Extension*) For each function $\mathbf{u}_\mathrm{C} \in \mathcal{H}(\mathbf{curl}, D_\mathrm{C})$, there exists exactly one weak divergence free function $\mathbf{u}_\mathrm{E} \in \mathcal{H}(\mathbf{curl}, D_\mathrm{E})$ that is the weak solution of

$$\mathbf{curl}\,(\nu_0\, \mathbf{curl}\, \mathbf{u}_\mathrm{E}) = \mathbf{J}, \qquad \text{in } D_\mathrm{E}, \tag{3.7a}$$

$$\mathbf{u}_\mathrm{E} \times \mathbf{n} = 0, \qquad \text{on } \Gamma_\mathrm{D} \cap \partial D_\mathrm{E}, \tag{3.7b}$$

$$\mathbf{u}_\mathrm{E} \times \mathbf{n} = \mathbf{u}_\mathrm{C} \times \mathbf{n}, \qquad \text{on } \partial D_\mathrm{C} \cap \partial D_\mathrm{E}, \tag{3.7c}$$

where $\mathbf{n}$ is the outer unit normal vector to ∂D_E and $\mathbf{J}$ is divergence free. The associated mapping $\mathbf{u}_\mathrm{E} = \mathcal{U}(\mathbf{u}_\mathrm{C})$ is called the (curl)-harmonic extension operator.

The space $\mathcal{H}(\mathbf{curl}, D)$ consists of square-integrable functions with (weak) square-integrable **curl** on D. Based on the previous definition we can define the function space for the weak formulation as

$$W(D) := \{\mathbf{u} \in \mathcal{H}(\mathbf{curl}, D) \mid \mathbf{u}_\mathrm{C} \in \mathcal{H}(\mathbf{curl}, D_\mathrm{C}), \mathbf{u}_\mathrm{E} = \mathcal{U}(\mathbf{u}_\mathrm{C}), \mathbf{u} \times \mathbf{n} = 0 \text{ on } \Gamma_\mathrm{D}\}. \tag{3.8}$$

To lighten notation, we introduce a vector function $\mathbf{h} : D \times \mathbb{R}^3 \to \mathbb{R}^3$ as

$$\mathbf{h}(\mathbf{x}, \mathbf{r}) := \nu(\mathbf{x}, |\mathbf{r}|)\mathbf{r}, \tag{3.9}$$

such that $\mathbf{h}(\mathbf{x}, \mathbf{B})$ represents the magnetic field strength, but with explicit dependence on $\mathbf{B}$. Then the weak formulation reads, almost everywhere in I_T, find $\mathbf{A}(t) \in W(D)$, $\dot{\mathbf{A}}|_{D_\mathrm{C}} \in L^2(D_\mathrm{C})^3$, subject to the initial condition, such that

$$(\sigma_\mathrm{C}\dot{\mathbf{A}}(t), \mathbf{v})_{D_\mathrm{C}} + (\mathbf{h}(\cdot, \mathbf{curl}\, \mathbf{A}(t)), \mathbf{curl}\, \mathbf{v})_D = (\mathbf{J}, \mathbf{v})_D, \quad \forall \mathbf{v} \in W(D). \tag{3.10}$$

Note that $\mathbf{A}$ is not assumed to be continuous in advance and the regularity of $\dot{\mathbf{A}}(t)|_{D_\mathrm{C}}$ could be further reduced [4]. We state the unique resolvability here, where the proof can be found in [4]. For a more detailed description of the Schur complement technique, see also [5].

Let $\mathcal{C}(X, Y)$ be the space of continuous functions from X into Y.

Theorem 3.1 *Let Assumptions 2.1, 2.2 and 3.1 be satisfied. Then the weak form of the magnetoquasistatic model (3.10) possesses a unique solution* $\mathbf{A} \in \mathcal{C}(I_T, W(D))$.

Before concluding this section, let us describe two simplified cases, that will be used frequently. If there is no conductive material present or temporal changes are sufficiently slow, the first term in (3.10) can be dropped and we recover the magnetostatic model. Here, we incorporate gauging into the functional space,

$$W_\mathrm{st}(D) := \{\mathbf{u} \in \mathcal{H}_0(\mathbf{curl}, D) \mid (\mathbf{u}, \mathbf{grad}\, \phi)_D = 0, \ \forall \phi \in \mathcal{H}_0(\mathbf{grad}, D)\}. \tag{3.11}$$

With a subscript 0 we denote a vanishing trace on the boundary ∂D, i.e., $\mathbf{u} \times \mathbf{n} = 0$ and $u = 0$ for $\mathcal{H}(\mathbf{curl}, D)$ and $\mathcal{H}(\mathbf{grad}, D)$, respectively. We endow $W_\mathrm{st}(D)$ with the norm $\|\mathbf{u}\|_{\mathcal{H}(\mathbf{curl}, D)} := \|\mathbf{curl}\, \mathbf{u}\|_2$. That this is indeed a norm follows from the Poincaré-Friedrich's inequality for $\mathbf{u} \in W_\mathrm{st}(D)$

$$\|\mathbf{u}\|_2 \leq C_\mathrm{F} \|\mathbf{curl}\, \mathbf{u}\|_2, \tag{3.12}$$

see [6, Corollary 3.19.]. We introduce

$$a(\mathbf{A}; \mathbf{v}) := (\mathbf{h}(\cdot, \mathbf{curl}\, \mathbf{A}), \mathbf{curl}\, \mathbf{v})_D, \tag{3.13a}$$

$$l(\mathbf{v}) := (\mathbf{J}, \mathbf{v})_D, \tag{3.13b}$$

where both a and f are linear in $\mathbf{v}$. Then the associated weak formulation reads, find $\mathbf{A} \in W_\mathrm{st}(D)$ such that

$$a(\mathbf{A};\mathbf{v}) = l(\mathbf{v}), \quad \forall \mathbf{v} \in W_{\mathrm{st}}(D). \tag{3.14}$$

For a divergence free current density $\mathbf{J}$, the strong form of (3.14) is given by

$$\begin{aligned}
\mathbf{curl}\ (\nu(\cdot, |\mathbf{curl}\ \mathbf{A}|)\mathbf{curl}\ \mathbf{A}) &= \mathbf{J}, && \text{in } D_{\mathrm{C}} \cup D_{\mathrm{E}}, && (3.15a)\\
\mathbf{n} \times (\mathbf{A}_{\mathrm{C}} - \mathbf{A}_{\mathrm{E}}) &= 0, && \text{on } \Gamma_{\mathrm{I}}, && (3.15b)\\
\mathbf{n} \times (\nu_{\mathrm{C}}\, \mathbf{curl}\ \mathbf{A}_{\mathrm{C}} - \nu_0\, \mathbf{curl}\ \mathbf{A}_{\mathrm{E}}) &= 0, && \text{on } \Gamma_{\mathrm{I}}, && (3.15c)\\
\mathbf{A} \times \mathbf{n} &= 0, && \text{on } \Gamma_{\mathrm{D}}, && (3.15d)\\
\operatorname{div} \mathbf{A} &= 0, && \text{in } D. && (3.15e)
\end{aligned}$$

Although the existence of a unique solution in the static case is assured by Theorem 3.1, we recall a more direct result, relying on tools that will be used later on. It is convenient to introduce an operator $\mathcal{M} : \mathcal{H}(\mathbf{curl}, D) \to \mathcal{H}(\mathbf{curl}, D)^*$ as $< \mathcal{M}\ \mathbf{A}, \mathbf{v} > := a(\mathbf{A}; \mathbf{v})_D$, with the following properties.

Lemma 3.1 *Let Assumption 2.2 be satisfied, there holds* $\forall \mathbf{u}, \mathbf{v} \in W_{\mathrm{st}}(D)$

$$< \mathcal{M}\ \mathbf{u}, \mathbf{u} - \mathbf{v} > - < \mathcal{M}\ \mathbf{v}, \mathbf{u} - \mathbf{v} > \geq \nu_{\min} \|\mathbf{u} - \mathbf{v}\|^2_{\mathcal{H}(\mathbf{curl}, D)}, \tag{3.16a}$$

$$| < \mathcal{M}\ \mathbf{u}, \mathbf{w} > - < \mathcal{M}\ \mathbf{v}, \mathbf{w} > | \leq 3\nu_0 \|\mathbf{u} - \mathbf{v}\|_{\mathcal{H}(\mathbf{curl}, D)} \|\mathbf{w}\|_{\mathcal{H}(\mathbf{curl}, D)}. \tag{3.16b}$$

Proof See, e.g., the Appendix of [7]. □

By means of the previous Lemma the operator $\mathcal{M}$ is strongly monotone and Lipschitz continuous on $W_{\mathrm{st}}(D)$. Then by the Theorem of Zarantonello [8, Theorem 25.B], the model (3.14) possesses a unique solution. Note that the divergence-free constraint can be also formulated explicitly by means of a mixed formulation: find $(\mathbf{A}, \lambda) \in \mathcal{H}_0(\mathbf{curl}, D) \times \mathcal{H}_0(\mathbf{grad}, D)$, such that

$$(\mathbf{h}(\cdot, \mathbf{curl}\ \mathbf{A}), \mathbf{curl}\ \mathbf{v})_D + (\mathbf{grad}\ \lambda, \mathbf{v})_D = (\mathbf{J}, \mathbf{v})_D, \tag{3.17}$$

$$(\mathbf{A}, \mathbf{grad}\ v)_D = 0, \tag{3.18}$$

for all $(\mathbf{v}, v) \in \mathcal{H}_0(\mathbf{curl}, D) \times \mathcal{H}_0(\mathbf{grad}, D)$, see also [9].

A second simplification relies on the observation that, e.g., in magnets or machines, the device configuration remains almost unchanged along one specific dimension. Then it is reasonable to assume that the field component in this direction is negligible and a two dimensional analysis can be carried out for cuts far away from the boundaries. This is also reasonable in preliminary design steps as three-dimensional design and optimization are still a computational challenge nowadays. Note that the two-dimensional modeling also addresses the important case of devices with axial symmetry. Consider a local (Cartesian) coordinate system $\mathbf{A} = (A_x, A_y, A_z)$ and the x-y plane: the curl operator reduces to $\mathbf{curl}((0, 0, A_z)) = (\partial_y A_z, -\partial_x A_z, 0)$ and we have $|\mathbf{curl}((0, 0, A_z))| = |\mathbf{grad}\ ((0, 0, A_z))|$. Setting $u = A_z$, the solution space is given by

$$W_{2\mathrm{D}}(D) := \mathcal{H}_0(\mathbf{grad}, D). \tag{3.19}$$

Then the two-dimensional equivalent of (3.10) reads, almost everywhere in I_T, find $u(t) \in W_{2\mathrm{D}}(D)$, $\dot{u}(t)|_{D_\mathrm{C}} \in L^2(D_\mathrm{C})$ such that

$$\int_D \sigma \frac{\partial u}{\partial t} v \,\mathrm{d}\mathbf{x} + \int_D \nu(\cdot, |\mathbf{grad}\ u|)\mathbf{grad}\ u \cdot \mathbf{grad}\ v \,\mathrm{d}\mathbf{x} = \int_D J_z v \,\mathrm{d}\mathbf{x}, \forall v \in W_{2\mathrm{D}}(D), \tag{3.20}$$

in particular there is no need for gauging anymore.

3.2 Reformulation as a Minimization Problem

In this section we consider the magnetostatic formulation and recall that it can be reformulated as a minimization problem. We will rely on notions from differential calculus and establish the differentiability of the operator $\mathcal{M}$. This will be frequently used later on for linearization and sensitivity analysis. We recall the following notions of derivatives from [10, p. 112]. Let X, Y be normed vector spaces, $U \subset X$ open and $g : U \to Y$.

Definition 3.3 (*Gâteaux Derivative*) We call g Gâteaux differentiable at $u \in U$ if there exists a continuous linear operator $g' : X \to Y$ such that

$$\lim_{s \to 0} (g(u + sv) - g(u))/s = g'(v). \tag{3.21}$$

Gâteaux differentiability is not sufficient if we want to apply, e.g., a first order Taylor expansion. In this case we employ the following derivative.

Definition 3.4 (*Fréchet Derivative*) We call g Fréchet differentiable at $u \in U$ if there exists a continuous linear operator $\delta g : X \to Y$ such that

$$g(u + v) = g(u) + \delta g(v) + o(\|v\|). \tag{3.22}$$

It is well known, that (3.14) can also be recast as the variational formulation,

$$F(\mathbf{u}) := E(\mathbf{u}) - l(\mathbf{u}) = \min!, \quad \mathbf{u} \in W_{\mathrm{st}}(D) \tag{3.23}$$

see, e.g., [8], where E refers to the magnetic energy

$$E(\mathbf{A}) = \frac{1}{2} \int_D \int_0^{|\mathbf{curl}\ \mathbf{A}|^2} \nu(\mathbf{x}, s) \mathrm{d}s \,\mathrm{d}\mathbf{x}. \tag{3.24}$$

To establish the existence of a minimizer we have to study the first and second order Fréchet derivative of F. As this will be also needed to apply the Newton method and for sensitivity analysis we work out this point in some detail, starting with a definition of the differential reluctivity.

Definition 3.5 (*Differential Reluctivity*) The differential reluctivity ν_{d} is defined, for $s \in \mathbb{R}^+$, as

$$\nu_{\mathrm{d}}(\cdot, s) := \nu(\cdot, s) + \nu^{(1)}(\cdot, s)s. \tag{3.25}$$

Moreover, the differential reluctivity tensor $\boldsymbol{\nu}_{\mathrm{d}}$ is defined, for $\mathbf{r} \in \mathbb{R}^3$, as

$$\boldsymbol{\nu}_{\mathrm{d}}(\cdot, \mathbf{r}) = \nu(\cdot, |\mathbf{r}|)\,\mathbf{I} + \frac{{}^{\circ(1)}(\cdot, |\mathbf{r}|)}{|\mathbf{r}|}\mathbf{r} \otimes \mathbf{r}, \tag{3.26}$$

where $\mathbf{I}$ and $\otimes$ refer to the 3×3 identity matrix and the tensor product, respectively.

In the previous definition we have set $\nu^{(1)}(\cdot, s) := \partial_s \nu(\cdot, s)$. This definition makes sense, whenever f_{BH} satisfies (2.14) as in this case ν is differentiable [11]. The differential reluctivity arises when differentiating $\nu(\cdot, s)s$ with respect to s. Moreover, the differential reluctivity tensor can be identified with the Jacobian of $\mathbf{h}(\cdot, \mathbf{r})$, as stated the following Lemma:

Lemma 3.2 *The Jacobian matrix of* $\mathbf{h}(\cdot, \mathbf{r})$ *with respect to* $\mathbf{r}$ *is given by the differential reluctivity tensor* $D\mathbf{h} = \boldsymbol{\nu}_{\mathrm{d}}$.

Proof For $\mathbf{x} \in D_{\mathrm{E}}$ the result follows immediately. For $\mathbf{x} \in D_{\mathrm{C}}$ we have $\mathbf{h}(\mathbf{x}, \mathbf{r}) = \nu_{\mathrm{C}}(|\mathbf{r}|)\mathbf{r}$ and hence

$$(D\mathbf{h})_{ij} = \partial(\nu_{\mathrm{C}}(|\mathbf{r}|)r_j)/\partial r_i = \nu_{\mathrm{C}}(|\mathbf{r}|)\delta_{ij} + \frac{\nu_{\mathrm{C}}^{(1)}(|\mathbf{r}|)}{|\mathbf{r}|}r_j r_i, \tag{3.27}$$

where δ_{ij} refers to the Kronecker delta and the result follows. □

We assume that the differential reluctivity satisfies the following properties.

Assumption 3.2 Both ν and $\boldsymbol{\nu}_{\mathrm{d}}$ are continuously differentiable. Moreover,

$$|\boldsymbol{\nu}_{\mathrm{d}}(\cdot, \mathbf{r})| \leq 3\nu_0, \tag{3.28}$$

$$\mathbf{r}^\top \boldsymbol{\nu}_{\mathrm{d}}(\cdot, \mathbf{r})\mathbf{r} \geq \nu_{\min}|\mathbf{r}|^2, \tag{3.29}$$

holds for all $\mathbf{r} \in \mathbb{R}^3$.

It can be shown, see again [11], that this is satisfied provided f_{BH} satisfies (2.14). Now we define, for $\mathbf{w} \in \mathcal{H}(\mathbf{curl}, D)$, the bilinear form $a(\mathbf{w}; \cdot, \cdot)' : \mathcal{H}(\mathbf{curl}, D) \times \mathcal{H}(\mathbf{curl}, D) \to \mathbb{R}$ as

$$a'(\mathbf{w}; \mathbf{u}, \mathbf{v}) := (\boldsymbol{\nu}_{\mathrm{d}}(\cdot, \mathbf{curl}\ \mathbf{w})\mathbf{curl}\ \mathbf{u}, \mathbf{curl}\ \mathbf{v})_D. \tag{3.30}$$

Note that $a'(\mathbf{w};\cdot,\cdot)$ is nonlinear in the first argument and by Lemma (3.2) equivalent to the Jacobian matrix of a at $\mathbf{w}$, if evaluated on discrete subspaces. Important properties of a' are stated as follows.

Lemma 3.3 *Under Assumption 3.2, the bilinear form $a'(\mathbf{w},\cdot,\cdot)$ is continuous and coercive on $W_{\mathrm{st}}(D)$, i.e., for all $\mathbf{u},\mathbf{v}\in W_{\mathrm{st}}(D)$, we have*

$$|a'(\mathbf{w};\mathbf{u},\mathbf{v})| \le C_1\|\mathbf{u}\|_{\mathcal{H}(\mathbf{curl},D)}\|\mathbf{v}\|_{\mathcal{H}(\mathbf{curl},D)}, \tag{3.31a}$$

$$a'(\mathbf{w};\mathbf{u},\mathbf{u}) \ge C_2\|\mathbf{u}\|^2_{\mathcal{H}(\mathbf{curl},D)}. \tag{3.31b}$$

Proof This has been established in the two-dimensional case in [11] and is readily extended to our setting. □

Now all required tools are at hand. A necessary condition for a minimum is that the Fréchet derivative of F vanishes. This gives rise to the Euler-Lagrange equation (3.14). Moreover, by Lemma 3.3, F is convex and hence the minimum is global [11, p. 26].

3.3 Space Discretization

Many different, well-established, space discretization schemes for Maxwell's equations are at our disposal to be applied to (3.10). We mention in particular finite elements [12, 13], the finite integration technique [14] and boundary element methods [15]. Here, we will make use of the finite element method due to its geometric flexibility and its capabilities of treating nonlinearities. A major issue with regard to the discretization of electromagnetic laws are the structural *topological* properties, that need to be preserved on the discrete level. This is successfully addressed within finite difference or finite volume like cochain approximations [16, 17] and discrete differential forms, e.g., Whitney forms, in the finite element context [18, 19]. In the latter case the key feature is that the discrete subspaces form a sub-complex of the de Rham or Hilbert complex [20, 21] with a commuting property. Establishing a subdivision $\mathcal{T}$ of the computational domain D, is the first important step for a finite element discretization. It consists of a finite collection of closed subsets as defined, e.g., in [22, p. 82]. As we are going to work with different types of subdivisions they will be defined in the respective context. A common assumption is that the mesh is aligned at the material interface. Let $h>0$ denote the maximum mesh size, serving as index for a family of subdivisions $\mathcal{T}_h$. For the time being we assume that $\mathcal{T}_h$ is quasi-uniform, i.e., $h/\mathrm{diam}(K)$ is bounded for all $K\in\mathcal{T}_h$, see [22, p. 120] for a precise definition. With $\mathcal{T}_h$ at hand, finite dimensional subspaces of $\mathcal{H}(\mathbf{grad},D)$, $\mathcal{H}(\mathbf{curl},D)$, $\mathcal{H}(\mathrm{div},D)$ and $L^2(D)$, denoted with subscript h, are defined in such a way that we obtain a commuting diagram, see Table 3.1. Projection operators from the continuous into the discrete spaces are denoted $P_h^{\mathbf{grad}}$, $P_h^{\mathbf{curl}}$, P_h^{div}, P_h, respectively. We refer to

Table 3.1 Commuting de Rham diagram without boundary conditions for the numerical approximation of electromagnetic laws. Discrete representatives are obtained via projection from the continuous spaces, which commutes with the respective differential operators

$$\begin{array}{ccccccccccc}
\mathbb{R} & \rightarrow & \mathcal{H}(\mathbf{grad}, D) & \xrightarrow{\mathbf{grad}} & \mathcal{H}(\mathbf{curl}, D) & \xrightarrow{\mathbf{curl}} & \mathcal{H}(\mathrm{div}, D) & \xrightarrow{\mathrm{div}} & L^2(D) & \rightarrow & 0 \\
 & & P_h^{\mathbf{grad}} \downarrow & & P_h^{\mathbf{curl}} \downarrow & & P_h^{\mathrm{div}} \downarrow & & P_h \downarrow & & \\
\mathbb{R} & \rightarrow & \mathcal{H}_h(\mathbf{grad}, D) & \xrightarrow{\mathbf{grad}} & \mathcal{H}_h(\mathbf{curl}, D) & \xrightarrow{\mathbf{curl}} & \mathcal{H}_h(\mathrm{div}, D) & \xrightarrow{\mathrm{div}} & L_h^2(D) & \rightarrow & 0
\end{array}$$

[20] for a construction for the important case of Whitney forms, see also Sect. 3.3.1. Note that these are not the *canonical* operators [20], based on point, edge and face degrees of freedom, as the continuous spaces involved fail to provide sufficient regularity. The commuting property refers to the fact that projection and application of the respective differential operator commute, e.g., $\mathbf{curl}\, P_h^{\mathbf{curl}} \mathbf{u} = P_h^{\mathrm{div}} \mathbf{curl}\, \mathbf{u}$, for $\mathbf{u} \in \mathcal{H}(\mathbf{curl}, D)$. Due to $\mathbf{curl}\,\mathbf{grad} = 0$ and $\mathrm{div}\,\mathbf{curl} = 0$ the image of a differential operator is a subset of the kernel of its successor in the diagram. Moreover, for simply connected domains the diagram is exact, i.e., both sets coincide, see, e.g., [20]. Let us give two examples of discrete spaces, which will be used in this work. Thereby, we will only give a short definition since both have been illustrated thoroughly in the literature.

3.3.1 Higher Order Whitney Forms

For $\mathcal{T}_h$, referring to a conformal tetrahedral mesh, several sub-complexes, based on piecewise polynomial spaces have been proposed. A rather general and unified exposition has been given in [20] based on polynomial spaces of differential forms. For a tetrahedron $K \in \mathcal{T}_h$, let $\mathcal{P}_q(K)$ and $\mathcal{P}_q(K)^3$ denote the space of polynomials of degree q on K and its vectorial counterpart, respectively. The definition of edge elements relies on the subspace $S_q(K)$ of $\mathcal{P}_q(K)^3$, consisting of polynomials such that $\mathbf{s}(\mathbf{x}) \cdot \mathbf{x} = 0$, $\forall \mathbf{x} \in K$. Then, we recall the finite elements

$$\mathcal{H}_h(\mathbf{grad}, D) := \{u \in \mathcal{H}(\mathbf{grad}, D) \mid u \in \mathcal{P}_q(K),\ \forall K \in \mathcal{T}_h\}, \tag{3.32a}$$

$$\mathcal{H}_h(\mathbf{curl}, D) := \{\mathbf{u} \in \mathcal{H}(\mathbf{curl}, D) \mid \mathbf{u} \in \mathcal{P}_{q-1}(K)^3 + S_q(K),\ \forall K \in \mathcal{T}_h\}, \tag{3.32b}$$

$$\mathcal{H}_h(\mathrm{div}, D) := \{\mathbf{v} \in \mathcal{H}(\mathrm{div}, D) \mid \mathbf{v} \in \mathcal{P}_{q-1}(K)^3 + \mathbf{x}\mathcal{P}_{q-1}(K),\ \forall K \in \mathcal{T}_h\}, \tag{3.32c}$$

$$L_h^2(D) := \{\varphi \in L^2(D) \mid \varphi \in \mathcal{P}_{q-1}(K),\ \forall K \in \mathcal{T}_h\}. \tag{3.32d}$$

In (3.32), $\mathcal{H}(\mathrm{div}, D)$ refers to the space of functions $\mathbf{u} \in L^2(D)^3$ with weak divergence $\mathrm{div}\,\mathbf{u} \in L^2(D)$. For the case $q = 1$ we recover the well-known lowest order Whitney forms. Finite elements in (3.32) are also referred to as continuous Lagrange elements of degree q, $\mathcal{H}(\mathbf{curl})$- and $\mathcal{H}(\text{div})$-conforming Nédélec elements of the first kind [18] of degree $q-1$, as well as discontinuous elements of degree $q-1$, respectively. We omit specifying the respective degrees of freedoms, implied by the continuity requirements of the continuous spaces and again refer to the literature, e.g., [18, 20].

3.3.2 Spline Finite Elements

Meshing of CAD geometries is a thoroughly investigated, though sometimes costly procedure, in particular in the context of optimization. In recent years much effort has been devoted to overcome this issue by directly working with spline based shape representations on the finite element level in the context of isogeometric analysis [23]. More generally, isogeometric analysis, as introduced in [23], aims at closing the gap between CAD and finite element analysis tools. Following the isoparametric concept, splines are both employed for approximating the solution of the partial differential equations and for representing the geometry.

These techniques have been extended to electromagnetics in [24] and further developed in [25, 26]. Let the computational domain be parametrized by a mapping $\mathbf{F}: \hat{D} \to D$, where $\hat{D}$ is the unit cube. Details on $\mathbf{F}$, typically realized by means of B-splines or Non-Uniform Rational B-Splines (NURBS), will be given in Sect. 4.2.3. Now, discrete spaces are first constructed on $\hat{D}$ and then *pushed forward* to D. To this end, a mesh on $\hat{D}$ is obtained as the Cartesian product of three partitions $\Pi_{x,y,z}$ of $[0,1]$. In this context $\mathcal{T}_h$ is defined by applying $\mathbf{F}$ to $\Pi_x \times \Pi_y \times \Pi_z$. We introduce the N-dimensional spline space $\mathcal{S}_N^{q,k}$, with functions that are piecewise polynomials of degree q and k-times continuously differentiable at each knot/point of the partition $\Pi_{x,y,z}$. For a precise definition of B-splines see Appendix B. By means of the tensor product B-spline space $\mathcal{S}^{q_1,q_2,q_3}_{\mathbf{k}_1,\mathbf{k}_2,\mathbf{k}_3} := \mathcal{S}^{q_1,\mathbf{k}_1} \otimes \mathcal{S}^{q_2,\mathbf{k}_2} \otimes \mathcal{S}^{q_3,\mathbf{k}_3}$ we define

$$\mathcal{H}_h(\mathbf{grad}, D) := \{u \mid \mathbf{F}^*(u) \in \mathcal{S}^{q_1,q_2,q_3}_{\mathbf{k}_1,\mathbf{k}_2,\mathbf{k}_3}\}, \tag{3.33a}$$

$$\mathcal{H}_h(\mathbf{curl}, D) := \{\mathbf{u} \mid \mathbf{F}^*(\mathbf{u}) \in \mathcal{S}^{q_1-1,q_2,q_3}_{\mathbf{k}_1-1,\mathbf{k}_2,\mathbf{k}_3} \otimes \mathcal{S}^{q_1,q_2-1,q_3}_{\mathbf{k}_1,\mathbf{k}_2-1,\mathbf{k}_3} \otimes \mathcal{S}^{q_1,q_2,q_3-1}_{\mathbf{k}_1,\mathbf{k}_2,\mathbf{k}_3-1}\}, \tag{3.33b}$$

$$\mathcal{H}_h(\text{div}, D) := \{\mathbf{v} \mid \mathbf{F}^*(\mathbf{v}) \in \mathcal{S}^{q_1,q_2-1,q_3-1}_{\mathbf{k}_1,\mathbf{k}_2-1,\mathbf{k}_3-1} \otimes \mathcal{S}^{q_1-1,q_2,q_3-1}_{\mathbf{k}_1-1,\mathbf{k}_2,\mathbf{k}_3-1} \otimes \mathcal{S}^{q_1-1,q_2-1,q_3}_{\mathbf{k}_1-1,\mathbf{k}_2-1,\mathbf{k}_3}\}, \tag{3.33c}$$

$$L_h^2(D) := \{\varphi \mid \mathbf{F}^*(\varphi) \in \mathcal{S}^{q_1-1,q_2-1,q_3-1}_{\mathbf{k}_1-1,\mathbf{k}_2-1,\mathbf{k}_3-1}\}, \tag{3.33d}$$

see [25]. In (3.33), the *pullback* $\mathbf{F}^*$ is given by

$$u \circ \mathbf{F},\ D\mathbf{F}^\top(\mathbf{u} \circ \mathbf{F}),\ \det(D\mathbf{F})(D\mathbf{F})^{-1}(\mathbf{v} \circ \mathbf{F}),\ \det(D\mathbf{F})(\varphi \circ \mathbf{F}), \tag{3.34}$$

respectively, where in turn $D\mathbf{F}$ refers to the Jacobian of the transformation $\mathbf{F}$ and $\det(D\mathbf{F})$ to its determinant.

3.3.3 Finite Element Formulation

Given discrete spaces, a finite element formulation can be derived in a straightforward way by simply substituting them for their continuous counterparts. Let us start with the static formulation (3.14) and postpone the technically more involved time-transient case. We define

$$W_{\mathrm{st},h}(D) := \{\mathbf{u}_h \in \mathcal{H}_h(\mathbf{curl}, D) \mid (\mathbf{u}_h, \mathbf{v}_h)_D = 0, \ \forall \mathbf{v}_h \in G_h(D)\}, \tag{3.35}$$

where $G_h(D)$ in turn is obtained by replacing $\mathcal{H}(\mathbf{grad}, D)$ by $\mathcal{H}_h(\mathbf{grad}, D)$ in (3.3). Then the finite element formulation reads, find $\mathbf{A}_h \in W_{\mathrm{st},h}(D)$ such that

$$a(\mathbf{A}_h; \mathbf{v}_h) = l(\mathbf{v}_h), \quad \forall \mathbf{v}_h \in W_{\mathrm{st},h}(D). \tag{3.36}$$

The previous relation is in fact equivalent to a linear system of equations. Indeed, as $W_{\mathrm{st},h}(D)$ is spanned by shape functions $(\varphi_i^{\mathrm{e}})_{i=1}^{N_{\mathrm{e}}}$, where e refers to edge, we can set

$$K_{ij}(\mathbf{a}) := (\nu(\cdot, |\mathbf{curl}\,\mathbf{A}_h|)\mathbf{curl}\,\varphi_i^{\mathrm{e}}, \mathbf{curl}\,\varphi_j^{\mathrm{e}})_D, \tag{3.37}$$

$$j_i := l(\varphi_i^{\mathrm{e}}), \tag{3.38}$$

for $i, j = 1, \ldots, N_{\mathrm{e}}$. In (3.37), $\mathbf{a}$ refers to the edge based degrees of freedom associated to $\mathbf{A}_h$. They are subject to the nonlinear system of equations

$$\mathbf{K}(\mathbf{a})\mathbf{a} = \mathbf{j}. \tag{3.39}$$

We assume that N_{e} refers to the number of active edges, i.e., to all edges not lying on the boundary. Still the system of equations (3.39) is singular, as the gauging condition is not yet imposed.

The discrete counterpart of the mixed formulation (3.18) in turn has the algebraic representation

$$\begin{pmatrix} \mathbf{K}(\mathbf{a}) & \mathbf{B}^\top \\ \mathbf{B} & \mathbf{O} \end{pmatrix} \begin{pmatrix} \mathbf{a} \\ \lambda \end{pmatrix} = \begin{pmatrix} \mathbf{j} \\ \mathbf{0} \end{pmatrix}, \tag{3.40}$$

where $\mathrm{B}_{ij} := (\mathbf{grad}\,\varphi_i^{\mathrm{n}}, \varphi_j^{\mathrm{e}})_D$ and $(\varphi_i^{\mathrm{n}})_{i=1}^{N_{\mathrm{N}}}$ are the nodal shape functions and by $\mathbf{O}$ we denote the $N_{\mathrm{N}} \times N_{\mathrm{N}}$ matrix with all zero entries. Here, again, we only consider active nodes, i.e., all nodes on ∂D are removed. The solution of saddle point systems, such as (3.40) after linearization, is well understood and we refer to [27] and the references therein for appropriate algorithms.

Remark 3.1 In general, for the evaluation of $\mathbf{K}, \mathbf{B}, \mathbf{j}$, one needs to resort to numerical integration. The associated error can be investigated by means of the Strang Lemma. This has been done, e.g., for the nonlinear case in [11, 28]. Also, in a more complete treatment of the numerical approximation errors, the error for solving the nonlinear system of equations (3.39) should be considered.

3.3.4 Finite Element a Priori Error Analysis

For completeness, the main points of the finite element error analysis will be described in the following. Our starting point will be (3.36), although similar results for the mixed formulation can be given, e.g., in the context of the Hodge-Laplacian [20]. Here, a first complication arises from the fact that $W_{\mathrm{st},h}(D) \not\subset W_{\mathrm{st}}(D)$, since a discrete divergence free function is not divergence free in general. However, this difficulty can be overcome, if a discrete Poincaré-Friedrichs-type inequality, similar to the continuous case can be established. To this end let us formalize the assumptions on the discrete subspaces

Assumption 3.3 (*Discrete de Rham Subcomplex*) The spaces $\mathcal{H}_h(\mathbf{grad}, D) \subset \mathcal{H}(\mathbf{grad}, D)$, $\mathcal{H}_h(\mathbf{curl}, D) \subset \mathcal{H}(\mathbf{curl}, D)$, $\mathcal{H}_h(\mathrm{div}, D) \subset \mathcal{H}(\mathrm{div}, D)$ and $L^2_h(D) \subset L^2(D)$ form a discrete subcomplex of the de Rham complex. There exist bounded projections $P_h^{\mathbf{grad}}$, $P_h^{\mathbf{curl}}$, P_h^{div}, P_h from the de Rham complex to the sub-complex and the diagram commutes. Moreover, a discrete Poincaré-Friedrichs inequality

$$\|\mathbf{u}_h\|_2 \le C\|\mathbf{curl}\,\mathbf{u}_h\|_2, \ \forall \mathbf{u}_h \in W_{\mathrm{st},h}(D), \tag{3.41}$$

holds, where the constant C in the previous expression is independent of h.

This assumption is satisfied by the methods presented in Sects. 3.3.1 and 3.3.2. The respective commuting diagrams have been established in [20, 25]. Concerning the Poincaré-Friedrichs inequality we refer to [29, p. 58] for the Whitney finite element method. Based on the same techniques the result holds true for the spline complex as well, by referring to the gap property, see again [25]. Based on Assumption 3.3, a unique solution of (3.36) exists, see, e.g., [7]. The next step is a nonlinear version of Céa's Lemma.

Lemma 3.4 (Céa) *Let Assumption 3.3 hold true and let* $\mathbf{J}$ *be weak divergence free in* D*, then*

$$\|\mathbf{A} - \mathbf{A}_h\|_{\mathcal{H}(\mathbf{curl},D)} \le C \inf_{\mathbf{v}_h \in W_{\mathrm{st},h}(D)} \|\mathbf{A} - \mathbf{v}_h\|_{\mathcal{H}(\mathbf{curl},D)}. \tag{3.42}$$

Proof Using the continuous and discrete saddle point formulation with test function $\mathbf{v}_h \in \mathcal{H}_h(\mathbf{curl}, D)$ we obtain

$$a(\mathbf{A}; \mathbf{v}_h) - a(\mathbf{A}_h; \mathbf{v}_h) = (\mathbf{v}_h, \mathbf{grad}(\lambda_h - \lambda))_D, \ \forall \mathbf{v}_h \in \mathcal{H}_h(\mathbf{curl}, D). \tag{3.43}$$

As $\mathbf{J}$ is weakly divergence free, we infer

$$(\mathbf{grad}\,\lambda, \mathbf{grad}\,\phi)_D = 0, \quad \forall \phi \in \mathcal{H}_0(\mathbf{grad}, D) \tag{3.44}$$

$$(\mathbf{grad}\,\lambda_h, \mathbf{grad}\,\phi_h)_D = 0 \quad \forall \phi_h \in \mathcal{H}_h(\mathbf{grad}, D) \cap \mathcal{H}_0(\mathbf{grad}, D). \tag{3.45}$$

Hence, both λ and λ_h vanish, as they solve a Laplace problem with vanishing Dirichlet boundary condition. We deduce the Galerkin orthogonality property

$$a(\mathbf{A}; \mathbf{v}_h) - a(\mathbf{A}_h; \mathbf{v}_h) = 0, \ \forall \mathbf{v}_h \in W_{\mathrm{st},h}(D), \tag{3.46}$$

since $W_{\mathrm{st},h}(D) \subset \mathcal{H}_h(\mathbf{curl}, D)$. In a next step we compute for any $\mathbf{v}_h \in W_{\mathrm{st},h}(D)$,

$$\|\mathbf{A} - \mathbf{A}_h\|^2_{\mathcal{H}(\mathbf{curl},D)} \underbrace{\leq}_{(3.16)} C_1(a(\mathbf{A}; \mathbf{A} - \mathbf{A}_h) - a(\mathbf{A}_h; \mathbf{A} - \mathbf{A}_h)) \tag{3.47}$$

$$\underbrace{\leq}_{(3.46)} C_1(a(\mathbf{A}; \mathbf{A} - \mathbf{v}_h) - a(\mathbf{A}_h; \mathbf{A} - \mathbf{v}_h)) \tag{3.48}$$

$$\underbrace{\leq}_{(3.16)} C_2\|\mathbf{A} - \mathbf{A}_h\|_{\mathcal{H}(\mathbf{curl},D)}\|\mathbf{A} - \mathbf{v}_h\|_{\mathcal{H}(\mathbf{curl},D)}, \tag{3.49}$$

where $C_1, C_2 > 0$. The statement follows since $\mathbf{v}_h$ is arbitrary. □

By choosing $\mathbf{v}_h = P_h^{\mathbf{curl}}\,\mathbf{A}$ in the previous Lemma, the actual convergence rate depends on the approximation accuracy of $\mathcal{H}(\mathbf{curl}, D)$ by means of $\mathcal{H}_h(\mathbf{curl}, D)$ and on the regularity of the solution $\mathbf{A}$. Assume, e.g., that $\mathbf{A} \in \mathcal{H}^s(\mathbf{curl}, D)$, with $s \in (0, 1]$, i.e., both $\mathbf{A}$ and $\mathbf{curl}\,\mathbf{A}$ are in the fractional order Sobolev space $\mathcal{H}^s(D)^3$. Then, e.g., for the spline case and a sufficiently smooth mapping $\mathbf{F}$ (see [25] for details) we obtain

$$\|\mathbf{A} - \mathbf{A}_h\|_{\mathcal{H}(\mathbf{curl},D)} \leq Ch^s \left(\|\mathbf{A}\|_{\mathcal{H}^s(D)^3} + \|\mathbf{curl}\,\mathbf{A}\|_{\mathcal{H}^s(D)^3}\right). \tag{3.50}$$

Similar result can be established for other approximation spaces as well. We also refer to [3, 6] for a definition of fractional order Sobolev spaces and embedding results, respectively.

3.4 Linearization

Even after discretization, equation (3.14) cannot be solved directly due to the non-linearity. Therefore, it is approximated by a linear model, that, for the time being, following [1], is abstractly formulated as: find $\mathbf{A}_{\mathrm{L}} \in W_{\mathrm{st}}(D)$, such that

$$\mathbf{h}_{\mathrm{L}}(\cdot, \mathbf{curl}\,\mathbf{A}_{\mathrm{L}}), \mathbf{curl}\,\mathbf{v})_D = (\mathbf{J}, \mathbf{v}) \quad \forall \mathbf{v} \in W_{\mathrm{st}}(D), \tag{3.51}$$

where L refers to linearization. Note that the linearization point, generally denoted as $\mathbf{A}_{\mathrm{L},0}$, is not indicated explicitly in (3.51). In practice the linearized model is iterated until a sufficient accuracy has been obtained, as described in Algorithm A.1 in Appendix A.

Two very important examples, used frequently in this work, can be written as (3.51). Perhaps the most simple choice for a linearized model is given by

$$\mathbf{h}_{\mathrm{L}}(\cdot, \mathbf{curl}\, \mathbf{A}_{\mathrm{L}}) = \underbrace{\nu(\cdot, |\mathbf{curl}\, \mathbf{A}_{\mathrm{L},0}|)}_{=:\nu_{\mathrm{L},0}}\, \mathbf{curl}\, \mathbf{A}_{\mathrm{L}}. \tag{3.52}$$

Applying Algorithm 2 together with (3.52) is referred to as successive substitution or Picard-iteration here. It is evident in this case, that (3.51) is well-posed by the Lax-Milgram Lemma, as $\nu_{\mathrm{L},0} \in L^{\infty}(D)$ and $\nu_{\mathrm{L},0} \geq \nu_{\min} > 0$. The implementation is simple and the method robust, however, as stated in Lemma 3.5 below, the speed of convergence is typically not very high. Therefore, it might be more efficient to use Newton's method for linearization. It consists in setting

$$\mathbf{h}_{\mathrm{L}}(\cdot, \mathbf{curl}\, \mathbf{A}_{\mathrm{L}}) = \underbrace{\boldsymbol{\nu}_{\mathrm{d}}(\cdot, \mathbf{curl}\, \mathbf{A}_{\mathrm{L},0})}_{=:\boldsymbol{\nu}_{\mathrm{d},0}}\, \mathbf{curl}\, \mathbf{A}_{\mathrm{L}} + \mathbf{H}_{\mathrm{CO}}, \tag{3.53}$$

with the coercive magnetic field strength

$$\mathbf{H}_{\mathrm{CO}} := (\nu_{\mathrm{L},0} - \boldsymbol{\nu}_{\mathrm{d},0})\mathbf{curl}\, \mathbf{A}_{\mathrm{L},0}. \tag{3.54}$$

Although expression (3.53) is actually affine, we adapt the term linearization, following [30]. Iterating as described in Algorithm A.1 yields the well-known Newton-Raphson method. Again (3.51) is well-posed for every $\mathbf{A}_{\mathrm{L},0}$ with the choice (3.53) by Lemma 3.3 and the fact that $\|\cdot\|_{\mathcal{H}(\mathbf{curl},D)}$ is a norm on $W_{\mathrm{st}}(D)$. Finally, we are going to comment on convergence rates for the choices (3.52) and (3.53). For a definition of linear, superlinear and quadratic convergence see Appendix A.

Lemma 3.5 *For an initial point $\mathbf{A}_{\mathrm{L},0} \in W_{\mathrm{st}}(D)$ sufficiently close to the solution, the l-th iterate $\mathbf{A}^{\{l\}}$, obtained by the method of successive substitution, converges linearly to $\mathbf{A}$, i.e.,*

$$\|\mathbf{A}^{\{l+1\}} - \mathbf{A}\|_{\mathcal{H}(\mathbf{curl},D)} \leq \sqrt{1-\beta}\|\mathbf{A}^{\{l\}} - \mathbf{A}\|_{\mathcal{H}(\mathbf{curl},D)}, \tag{3.55}$$

where $\beta = \nu_0/\nu_{\min}\alpha$ and $\alpha \leq \nu_{\min}/(3\nu_0)$.

Proof The proof given [11, pp. 52–54] can be extended to our three-dimensional setting in a straight forward way. □

As already mentioned, in the case of successive substitution, the convergence is rather slow, i.e., linear. In the case of the Newton-Raphson method, a higher convergence rate can be obtained. However, the convergence analysis is more involved and in particular no convergence can be shown in the present continuous setting before discretization, see [11, Sect. 4.1.2] and also [31]. Therefore, we discuss the convergence of Newton's method after discretization, i.e., for the linearized solution $\mathbf{A}_{\mathrm{L},h} \in W_{\mathrm{st},h}(D)$ subject to

$$(\mathbf{h}_{\mathrm{L}}(\cdot, \mathbf{curl}\, \mathbf{A}_{\mathrm{L},h}), \mathbf{curl}\, \mathbf{v}_h)_D = (\mathbf{J}, \mathbf{v}_h), \quad \forall \mathbf{v}_h \in W_{\mathrm{st},h}(D). \tag{3.56}$$

For the analysis of Newton's method we also introduce

$$< \mathcal{M}_{\mathrm{L},h}(\mathbf{u}_h)\mathbf{v}_h, \mathbf{w}_h > := a'(\mathbf{u}_h; \mathbf{v}_h, \mathbf{w}_h), \tag{3.57}$$

where $\mathcal{M}_{\mathrm{L},h}(\mathbf{u}_h)$ is a linear and bounded operator from $\mathcal{H}_h(\mathbf{curl}, D)$ to $\mathcal{H}_h(\mathbf{curl}, D)^*$. We state the following convergence result, see [11, Theorem 4.2].

Lemma 3.6 *For the choice $\alpha = 1$, given $\mathbf{A}_h^{\{1\}} \in W_{\mathrm{st},h}(D)$, the l-th iterate $\mathbf{A}_h^{\{l\}}$ obtained by the Newton-Raphson method converges locally and superlinearly to $\mathbf{A}_h$ if the mapping $\mathbf{u}_h \mapsto \mathcal{M}_{\mathrm{L},h}(\mathbf{u}_h)$ is continuous. If additionally $\mathbf{u}_h \mapsto \mathcal{M}_{\mathrm{L},h}(\mathbf{u}_h)$ is Lipschitz continuous the convergence is locally quadratic.*

The continuity and Lipschitz continuity in the previous result have been established in the two-dimensional case after discretization in [11]. It should be noted, that the constants involved depend on the discretization parameter h and these results do not hold true in the limit $h \to 0$. We do not see any reason why the proof cannot be extended to $3D$. However, as establishing this proof is a rather technical task and not in the main scope of this work, we do not further investigate this issue.

We have described two different kinds of approximations at this stage, namely linearization and discretization. Naturally the question arises, whether these commute, i.e., if the systems of equations obtained by discretization before linearization and linearization before discretization are identical. Fortunately, this holds true as can be seen by a direct computation, performed in [32, pp. 41–45] for the Newton-Raphson method. Hence, the (un-gauged) discrete and linearized systems take the form

$$\mathbf{K}(\mathbf{a}_{\mathrm{L},0})\mathbf{a}_{\mathrm{L}} = \mathbf{j}, \tag{3.58a}$$

$$D\mathbf{K}(\mathbf{a}_{\mathrm{L},0})\mathbf{a}_{\mathrm{L}} = \mathbf{j} - \mathbf{j}_{\mathrm{CO}}, \tag{3.58b}$$

where $\mathbf{j}_{\mathrm{CO}}$ is the discrete representation of $\mathbf{J}_{\mathrm{CO}} := \mathbf{curl}\, \mathbf{H}_{\mathrm{CO}}$ and $D\mathbf{K}$ the Jacobian matrix. Note that because of div $\mathbf{curl} = 0$, the source current remains divergence free. Often both iteration schemes are given in the more explicit form

$$\mathbf{K}(\mathbf{a}^{\{l\}})(\mathbf{a}^{\{l+1\}} - \mathbf{a}^{\{l\}}) = -\alpha(\mathbf{K}(\mathbf{a}^{\{l\}}) - \mathbf{j}), \tag{3.59}$$

$$D\mathbf{K}(\mathbf{a}^{\{l\}})(\mathbf{a}^{\{l+1\}} - \mathbf{a}^{\{l\}}) = -\alpha(\mathbf{K}(\mathbf{a}^{\{l\}}) - \mathbf{j}). \tag{3.60}$$

Remark 3.2 Each linear system of equations given in (3.58), written compactly as $\tilde{\mathbf{K}}\mathbf{a}_{\mathrm{L}} = \tilde{\mathbf{j}}$, can be regularized by means of a mixed formulation as outlined in Sect. 3.3. A different approach is based on the preconditioned conjugate gradient (pcg) algorithm. It has been observed in [14, 33] that the pcg algorithm converges to

$$\mathbf{a}_{\mathrm{L}}^{\mathrm{end}} = \tilde{\mathbf{K}}^{+}\tilde{\mathbf{j}}, \tag{3.61}$$

where $\tilde{\mathbf{K}}^+$ denotes the pseudo-inverse of $\tilde{\mathbf{K}}$, provided that both the starting vector $\mathbf{a}_\mathrm{L}^\mathrm{init}$ associated to $\mathbf{A}_\mathrm{L}^\mathrm{init}$ and the source current $\tilde{\mathbf{j}}$ associated to $\tilde{\mathbf{J}}$ are discrete divergence free in a weak sense. We recall that this is expressed $\forall \mathbf{v}_h \in G_h(D)$ as

$$(\mathbf{A}_\mathrm{L}^\mathrm{init}, \mathbf{v}_h) = 0, \quad (\tilde{\mathbf{J}}, \mathbf{v}_h) = 0, \tag{3.62}$$

referred to as weak gauging [14, 34] as no further explicit gauging is required.

3.5 A Posteriori Error Analysis of Linearization and Discretization Error

For simplicity we consider the case without damping in this section, i.e., $\alpha = 1$. The a posteriori error analysis of discretization and linearization errors is of significant practical importance. In particular, a stopping criterion for the iterative linearization procedure (Algorithm A.1, Appendix A) is based on the control of the linearization error, which in turn should be balanced with the discretization error to keep the cost minimal. Sophisticated approaches have been proposed to this end and ideas of [1], developed further in [30, 35], are extended here to the present setting. Let $\mathbf{A}_{\mathrm{L},h}$ be the solution of (3.56). Based on the observation, that a suitable error measure for strongly monotone, quasi-linear problems, such as (3.36), is based on the dual norm of $\mathbf{h}$ [30] we introduce

$$\mathrm{err}_{\mathrm{L},h} := \sup_{\substack{\mathbf{v}\in W_{st}(D)\\ \mathbf{v}\neq 0}} \frac{(\mathbf{h}(\cdot, \mathbf{curl}\,\mathbf{A}) - \mathbf{h}(\cdot, \mathbf{curl}\,\mathbf{A}_{\mathrm{L},h}), \mathbf{curl}\,\mathbf{v})_D}{\|\mathbf{curl}\,\mathbf{v}\|_2}, \tag{3.63}$$

to quantify the discretization and linearization error. Due to the strong monotonicity of $\mathcal{M}$, this yields control on the error in the norm

$$\nu_{\min} \|\mathbf{A} - \mathbf{A}_{\mathrm{L},h}\|_{\mathcal{H}(\mathbf{curl}, D)} \le \mathrm{err}_{\mathrm{L},h}, \tag{3.64}$$

too, however, this might yield rough estimates in some situations. Here, error estimation is based on the following result, that we adapt from [1, Theorem 2.1]. Similar to the analysis of Newton's method we employ a linearized operator $\mathcal{M}_\mathrm{L} : \mathcal{H}(\mathbf{curl}, D) \to \mathcal{H}(\mathbf{curl}, D)^*$, given by

$$< \mathcal{M}_\mathrm{L}\mathbf{v}, \mathbf{w} > := (\mathbf{h}_\mathrm{L}(\cdot, \mathbf{curl}\,\mathbf{v}), \mathbf{curl}\,\mathbf{w})_D. \tag{3.65}$$

Theorem 3.2 *Let $\mathcal{M}$ be strongly monotone, Lipschitz continuous and $\mathcal{M}\,0 = 0$ hold. Moreover, let $\mathcal{M}_\mathrm{L}$ be coercive and continuous on $W_\mathrm{st}(D)$, then*

$$\mathrm{err}_{\mathrm{L},h} \leq \underbrace{\|\mathbf{h}_{\mathrm{L}}(\cdot, \mathbf{curl}\,\mathbf{A}_{\mathrm{L}}) - \mathbf{h}_{\mathrm{L}}(\cdot, \mathbf{curl}\,\mathbf{A}_{\mathrm{L},h})\|_2}_{=:\mathrm{err}_h} + \underbrace{\|\mathbf{h}_{\mathrm{L}}(\cdot, \mathbf{curl}\,\mathbf{A}_{\mathrm{L},h}) - \mathbf{h}(\cdot, \mathbf{curl}\,\mathbf{A}_{\mathrm{L},h})\|_2}_{=:\mathrm{err}_{\mathrm{L}}} . \quad (3.66)$$

Proof We observe that

$$\mathrm{err}_{\mathrm{L},h} = \|\mathcal{M}\,\mathbf{A} - \mathcal{M}\,\mathbf{A}_{\mathrm{L},h}\|_*, \quad (3.67)$$

where we abbreviate $\|\cdot\|_* := \|\cdot\|_{\mathcal{H}(\mathbf{curl},D)^*}$. Adding $\mathcal{M}_{\mathrm{L}}\mathbf{A}_{\mathrm{L},h} - \mathcal{M}_{\mathrm{L}}\mathbf{A}_{\mathrm{L},h}$ and using the triangle inequality we obtain

$$\mathrm{err}_{\mathrm{L},h} \leq \|\mathcal{M}\,\mathbf{A} - \mathcal{M}_{\mathrm{L}}\mathbf{A}_{\mathrm{L},h}\|_* + \|\mathcal{M}_{\mathrm{L}}\mathbf{A}_{\mathrm{L},h} - \mathcal{M}\,\mathbf{A}_{\mathrm{L},h}\|_*. \quad (3.68)$$

The result follows by using the Cauchy-Schwarz inequality and by $\mathcal{M}\,\mathbf{A} = \mathcal{M}_{\mathrm{L}}\mathbf{A}_{\mathrm{L}}$. This in turn is inferred from

$$< \mathcal{M}\,\mathbf{A}, \mathbf{v} > = (\mathbf{J}, \mathbf{v})_D, \quad (3.69)$$

$$< \mathcal{M}_{\mathrm{L}}\mathbf{A}_{\mathrm{L}}, \mathbf{v} > = (\mathbf{J}, \mathbf{v})_D, \quad (3.70)$$

for all $\mathbf{v} \in W_{\mathrm{st}}(D)$. □

Computable estimators for the error contributions $\mathrm{err}_{\mathrm{L}}$, err_h, referred to as linearization and discretization error, respectively, will now be discussed in some detail. First we observe that $\mathrm{err}_{\mathrm{L}}$ is directly computable and takes the form

$$\mathrm{err}_{\mathrm{L}} = \begin{cases} \|(\nu_{\mathrm{L},0} - \nu(\cdot, |\mathbf{curl}\,\mathbf{A}_{\mathrm{L},h}|))\mathbf{curl}\,\mathbf{A}_{\mathrm{L},h}\|_2, \\ \|(\boldsymbol{\nu}_{\mathrm{d},0} - \nu(\cdot, |\mathbf{curl}\,\mathbf{A}_{\mathrm{L},h}|))\mathbf{curl}\,\mathbf{A}_{\mathrm{L},h} + \mathbf{H}_{\mathrm{CO}}\|_2, \end{cases} \quad (3.71)$$

for the Picard and Newton method, respectively. Of course both errors vanish on D_{E}. Now, for err_h we can apply any available discretization error estimator for linear elliptic problems, as it is based on the linearized system

$$\mathbf{curl}\ (\tilde{\boldsymbol{\nu}}\mathbf{curl}\,\mathbf{A}_{\mathrm{L}}) = \tilde{\mathbf{J}}, \quad (3.72a)$$

$$\tilde{\boldsymbol{\nu}}\mathbf{curl}\,\mathbf{A}_{\mathrm{L}} = \mathbf{H}, \quad (3.72b)$$

where $\tilde{\boldsymbol{\nu}}$ and $\tilde{\mathbf{J}}$ reflect the different choices of linearization as described in Sect. 3.4. In particular

$$\tilde{\boldsymbol{\nu}} = \begin{cases} \nu_{\mathrm{L},0}, \\ \boldsymbol{\nu}_{\mathrm{d},0} \end{cases} \quad (3.73)$$

for the choices of successive approximation and the Newton-Raphson method, respectively and hence, $\tilde{\boldsymbol{\nu}}$ is a tensor in the general case. Here, we employ the hypercircle method [36]. A Prager-Synge identity for problems related to Maxwell's equations was given in [37, Theorem 10].

Theorem 3.3 *Let* $\mathbf{v} \in \mathcal{H}_0(\mathbf{curl}, D)$ *and* $\mathbf{H}_{\mathrm{eq}} \in \mathcal{H}(\mathbf{curl}, D)$ *satisfy*

$$\mathbf{curl}\ \mathbf{H}_{\mathrm{eq}} = \tilde{\mathbf{J}}. \tag{3.74}$$

Then for $(\mathbf{A}_{\mathrm{L}}, \mathbf{H})$ *subject to (3.72) there holds*

$$\|\tilde{\boldsymbol{\nu}}^{1/2}\mathbf{curl}\ (\mathbf{A}_{\mathrm{L}} - \mathbf{v})\|_2^2 + \|\tilde{\boldsymbol{\nu}}^{-1/2}(\mathbf{H} - \mathbf{H}_{\mathrm{eq}})\|_2^2 \leq \|\tilde{\boldsymbol{\nu}}^{1/2}(\mathbf{curl}\ \mathbf{v} - \tilde{\boldsymbol{\nu}}^{-1}\mathbf{H}_{\mathrm{eq}})\|_2^2. \tag{3.75}$$

Applying Theorem 3.3 with $\mathbf{v} = \mathbf{A}_{\mathrm{L},h}$, the discretization error can be estimated by means of

$$\mathrm{err}_h \leq \underbrace{\|\tilde{\boldsymbol{\nu}}^{1/2}\|_{L^\infty(D,\mathbb{R}^3\times\mathbb{R}^3)}}_{=:C_{\tilde{\nu}}} \|\tilde{\boldsymbol{\nu}}^{1/2}\mathbf{curl}\ (\mathbf{A}_{\mathrm{L}} - \mathbf{A}_{\mathrm{L},h})\|_2 \tag{3.76}$$

$$\leq C_{\tilde{\nu}}\|\tilde{\boldsymbol{\nu}}^{1/2}(\mathbf{curl}\ \mathbf{A}_{\mathrm{L},h} - \tilde{\boldsymbol{\nu}}^{-1}\mathbf{H}_{\mathrm{eq}})\|_2. \tag{3.77}$$

Given a matrix norm $\|\cdot\|_{\mathbb{R}^3\times\mathbb{R}^3}$, and the uniform norm

$$\|\mathrm{u}\|_{\mathrm{L}^\infty(\mathrm{D})} := \mathrm{ess\ sup}_{\mathbf{x}\in\mathrm{D}}|\mathrm{u}| \tag{3.78}$$

with the essential supremum [3, p. 27], the norm in relation (3.76) is defined as

$$\|\tilde{\boldsymbol{\nu}}^{1/2}\|_{L^\infty(D,\mathbb{R}^3\times\mathbb{R}^3)} := \mathrm{ess\ sup}_{\mathbf{x}\in D}\|\tilde{\boldsymbol{\nu}}^{1/2}\|_{\mathbb{R}^3\times\mathbb{R}^3}. \tag{3.79}$$

The delicate point in an estimate such as (3.75) is the construction of a so-called equilibrated flux $\mathbf{H}_{\mathrm{eq}}$. To this end we employ a mixed finite element method directly. Note that this should be seen as a proof of principle, as for practical applications, this would be prohibitively costly and an efficient reconstruction procedure, as described, e.g., in [37], should be used. The mixed method reads, find $(\mathbf{H}_{\mathrm{eq},h}, \tilde{\mathbf{A}}_h) \in \mathcal{H}_h(\mathbf{curl}, D) \times L_h^2(D)^3$ such that

$$(\tilde{\boldsymbol{\nu}}^{-1}\mathbf{H}_{\mathrm{eq},h}, \mathbf{v}_h)_D - (\tilde{\mathbf{A}}_h, \mathbf{curl}\ \mathbf{v}_h)_D = 0, \tag{3.80a}$$

$$(\mathbf{curl}\ \mathbf{H}_{\mathrm{eq},h}, \mathbf{w}_h)_D = (\tilde{\mathbf{J}}, \mathbf{w}_h)_D, \tag{3.80b}$$

for all $(\mathbf{v}_h, \mathbf{w}_h) \in \mathcal{H}_h(\mathbf{curl}, D) \times L_h^2(D)^3$. In particular this implies $\mathbf{curl}\ \mathbf{H}_{\mathrm{eq},h} = P_h\tilde{\mathbf{J}}$, i.e., $\mathbf{H}_{\mathrm{eq},h}$ is an equilibrated flux only if $P_h\tilde{\mathbf{J}} = \tilde{\mathbf{J}}$. Techniques to overcome this difficulty have been presented in [37], here we simply observe that the error of using $\mathbf{H}_{\mathrm{eq},h}$, for general, smooth $\tilde{\mathbf{J}}$, is of higher order as shown in [38] for the two-dimensional case.

Remark 3.3 For the case of the Newton-Raphson method, the derivation has to be considered as formal, as we do not now if $\tilde{\boldsymbol{\nu}} = \nu_{\mathrm{d},0}$ is invertible at all. However, for the case $\tilde{\boldsymbol{\nu}} = \nu_{\mathrm{L},0}$ this is guaranteed as $\nu_{\mathrm{L},0} > 0$ holds.

Example 3.1 (p-Laplace) For illustration, we consider a well-known two-dimensional and static example originally proposed in [1], which in the present form is adapted from [30]. Let the computational domain be $D = D_C = (0, 1) \times (0, 1)$. The p-Laplacian, given by

$$\nu_C(s) = s^{p-2}, \tag{3.81}$$

is a frequently used prototype for quasilinear elliptic models [39]. In [40], we have studied this example in a stochastic setting by modeling p as a random parameter. We assume homogeneous Dirichlet boundary conditions and impose a constant excitation $J = 2$. Then, on D, the analytical solution of

$$-\operatorname{div}\ (\nu(|\mathbf{grad}\ u|)\mathbf{grad}\ u) = J \tag{3.82}$$

is given by

$$u(\mathbf{x}) = -\frac{p-1}{p}|\mathbf{x} - (0.5, 0.5)|^{p/(p-1)} + \frac{p-1}{p}\left(\frac{1}{2}\right)^{p/(p-1)}, \tag{3.83}$$

see [30, p. 2789]. As $\nu(0) = 0$ and ν is unbounded for $s \to \infty$ and $p > 2$, (3.81) violates the assumptions on the reluctivity. Still continuity and monotonicity results can be established in this more general setting, as was shown in [35]. The solution is sought in $\mathcal{H}_0(\mathbf{grad}, D)$ and we employ, as outlined in Sect. 3.3.1, lowest order $\mathcal{H}(\mathbf{grad}, D)$-conforming elements for the finite element approximation of (3.82). The approximation is carried out on a uniform triangulation, based on the code FEniCS [32]. In two dimensions, (3.80) reduces to a mixed formulation for the nonlinear Poisson problem and lowest order Brezzi-Douglas-Marini $\mathcal{H}(\text{div})$ conforming elements and discontinuous Lagrange elements are used for its approximation, respectively. Linearization is carried out by means of the Picard method (3.52) with damping. The linearization error is estimated by (3.71), whereas (3.77) is used for the discretization error with $C_{\tilde{\nu}} = 1$. Error estimators for err_L, err_h and $\text{err}_{\text{L},h}$ are denoted Δ_L, Δ_h and $\Delta_{\text{L},h}$, respectively. As the true error $\text{err}_{\text{L},h}$ defined in (3.68) is not computable we employ the upper bound

$$\text{err}^*_{\text{L},h} := \|\nu(|\mathbf{grad}\ u_{\text{L},h}|)\mathbf{grad}\ u_{\text{L},h} - \nu(|\mathbf{grad}\ u|)\mathbf{grad}\ u\|_2. \tag{3.84}$$

Results are reported in Fig. 3.1, where on the left all error estimators and the *true* error $\text{err}^*_{\text{L},h}$ are depicted with respect to the number of iterations. We observe, that the iteration procedure converges and the linearization error is negligible after eight iterations. The error estimators can be used for error balancing of discretization and linearization error in practice in order to keep the number of iterations as small as possible. On the right, the overall estimated error is compared to $\text{err}^*_{\text{L},h}$ and the true

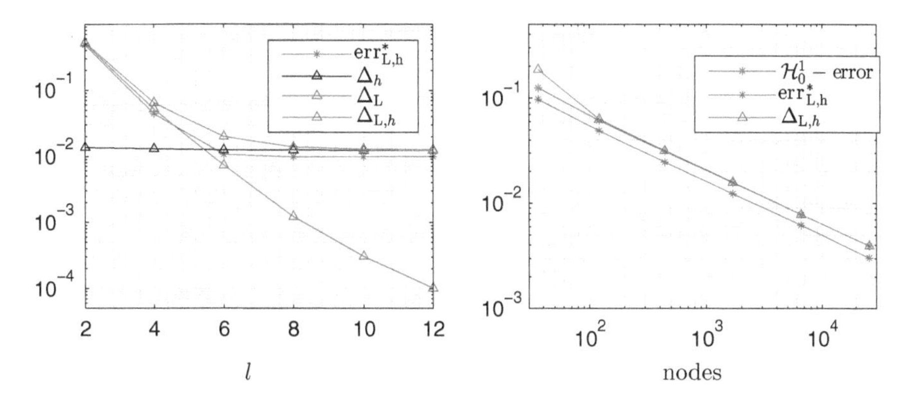

Fig. 3.1 Error estimation for the p-Laplace example. *Left* error estimation with respect to the number of iterations in comparison to an upper bound of the real error $\mathrm{err}^*_{\mathrm{L},h}$. Fixed uniform triangulation with 5000 triangles. *Right* error estimation with respect to the number of nodes in comparison to an upper bound of the real error $\mathrm{err}^*_{\mathrm{L},h}$ and the error in the $\mathcal{H}^1_0$-norm

$\mathcal{H}^1_0$-error with respect to the number of nodes. Linearization errors are controlled to be sufficiently small. We observe, that the error estimator is capable of accurately predicting the error for both $\mathrm{err}^*_{\mathrm{L},h}$ and the $\mathcal{H}^1_0$ norm.

3.6 Temporal Discretization

We now return to the time-dependent case and complement the description of the computational model by a time-discretization procedure. We denote the derivative of the time-dependent discrete magnetic potential vector $\mathbf{a}(t)$ as $\dot{\mathbf{a}}(t) := \mathrm{d}\mathbf{a}(t)/\mathrm{d}t$. Introducing the $N_\mathrm{e} \times N_\mathrm{e}$ mass matrix $M_{ij} := (\sigma_\mathrm{C}\varphi^\mathrm{e}_i, \varphi^\mathrm{e}_j)_{D_\mathrm{C}}$, the semi-discrete equations under consideration can be rewritten as

$$\mathbf{M}\dot{\mathbf{a}}(t) + \mathbf{K}(\mathbf{a}(t))\mathbf{a}(t) = \mathbf{j}, \tag{3.85}$$

for $t \in (0, T]$, $\mathbf{a}(0) = \mathbf{a}_{\mathrm{init}}$. As $\mathbf{M}$ is singular, the initial value problem (3.85) is of differential-algebraic type, which complicates its numerical approximation [41, 42]. A popular remedy is to introduce a small, artificial conductivity in the non-conducting region, such that zero eigenvalues in the matrix (3.85) are removed. The associated modeling error is well understood [2], however, we do not follow this approach here. We proceed by introducing a regular subdivision $\mathcal{T}_T := \{t_n \mid t_{n+1} > t_n,\ n = 1, \ldots, N_T - 1\}$ of $[0, T]$, with mesh size $h_n := t_{n-1} - t_n$ and maximum mesh size h_T. Applying a time-stepping procedure to (3.85) always involves implicit components, due to the algebraic part. Also the differential part is known to be stiff, giving rise to the problem of numerical stability for explicit schemes. Principally every time-stepping scheme that is well suited for stiff ordinary differential equations can be adapted here. For simplicity we focus on the implicit Euler method and obtain

$$(\mathbf{M} + h_T \mathbf{K}(\mathbf{a}_n))\, \mathbf{a}_n = h_T \mathbf{j}_n + \mathbf{M}\mathbf{a}_{n-1}, \quad \mathbf{a}_1 = \mathbf{a}_{\text{init}}, \tag{3.86}$$

for $n = 2, \ldots, N_T$, where $\mathbf{a}_n \approx \mathbf{a}(t_n)$. Note that the initial condition has to fulfill a consistency condition due to the algebraic part. Numerical methods involved in the approximation (3.86) are rather standard and their stability properties are well understood, see, e.g., [43, 44]. However, a complete approximation result establishing the convergence of the solution $\mathbf{A}_{h,h_T}$ of the fully discrete scheme (3.86) towards the weak solution (3.10) is apparently missing in the literature. In [45], for a slightly different current excitation and lowest order finite elements, the estimates

$$\|\mathbf{A} - \mathbf{A}_{h,h_T}\|_{L^2([0,T],W(D))} \leq C_1 h_T^{1/2} + C_2 h^{1/2} + C_3 h_T^{1/2} h^{-1/2}, \tag{3.87a}$$

$$\|\mathbf{A} - \mathbf{A}_{h,h_T}\|_{L^\infty([0,T],L^2(D_C)^3)} \leq C_1 h_T^{1/2} + C_2 h^{1/2} + C_3 h_T^{1/2} h^{-1/2}, \tag{3.87b}$$

were established for the case of a linear magnetic material. By investigating the proof, we claim that these results hold also true in the nonlinear case, by replacing the coercivity of the bilinear form by the strong monotonicity property. Estimate (3.87) is weaker than related results for purely parabolic problems due to a reduced regularity for solutions of mixed elliptic-parabolic equations [45]. More precisely, in order to achieve convergence, the refinement in the temporal variable must exceed the refinement in the spatial variable. However, the sharpness of (3.87) has not been confirmed by numerical experiments so far. Contributions for the linearization error could be included by the techniques outlined in Sect. 3.4.

3.7 Conclusion

This section was concerned with a weak formulation of the magnetoquasistatic model and a review of existing numerical techniques for its approximation. We presented finite element schemes for discretization in physical space, linearization methods and the implicit Euler method for discretization in time. For each approximation step, the associated a priori error analysis was sketched. Moreover, an a posteriori error analysis of linearization and finite element discretization error was outlined.

References

1. Chaillou, A.L., Suri, M.: Computable error estimators for the approximation of nonlinear problems by linearized models. Comput. Methods Appl. Mech. Eng. **196**(1), 210–224 (2006)
2. Bachinger, F., Langer, U., Schöberl, J.: Numerical analysis of nonlinear multiharmonic eddy current problems. Numerische Mathematik **100**(4), 593–616 (2005)
3. Adams, R.A., Fournier, J.J.F.: Sobolev spaces, vol. 140. Academic Press (2003)

4. Bachinger, F., Langer, U., Schöberl, J.: Numerical analysis of nonlinear multiharmonic eddy current problems. Technical report, SFB Numerical and Symbolic Scientific Computing, Johannes Kepler University Linz, Austria (2004)
5. Kolmbauer, M.: Existence and uniqueness of eddy current problems in bounded and unbounded domains. Technical Report 03, Johannes Kepler University Linz, Austria (2011)
6. Amrouche, C., Bernardi, C., Dauge, M., Girault, V.: Vector potentials in three-dimensional non-smooth domains. Math. Methods Appl. Sci. **21**(9), 823–864 (1998)
7. Yousept, I.: Optimal control of quasilinear $H(curl)$-elliptic partial differential equations in magnetostatic field problems. SIAM J. Control Optim. **51**(5), 3624–3651 (2013)
8. Zeidler, E.: Nonlinear functional analysis and its applications, II/B: Nonlinear monotone operators. Springer, New York (1990)
9. Monk, P.: Superconvergence of finite element approximations to Maxwell's equations. Numer. Methods Partial Differ. Equ. **10**(6), 793–812 (1994)
10. Werner, D.: Funktionalanalysis, vol. 2. Springer (2005)
11. Pechstein, C.: Multigrid-Newton-methods for nonlinear magnetostatic problems. M.Sc. thesis, Johannes Kepler Universität Linz, Austria (2004)
12. Biro, O., Preis, K.: On the use of the magnetic vector potential in the finite-element analysis of three-dimensional eddy currents. IEEE Trans. Magn. **25**(4), 3145–3159 (1989)
13. Bossavit, A.: On the numerical analysis of eddy-current problems. Comput. Methods Appl. Mech. Eng. **27**(3), 303–318 (1981)
14. Clemens, M., Weiland, T.: Transient eddy-current calculation with the FI-method. IEEE Trans. Magn. **35**(3), 1163–1166 (1999)
15. Mayergoyz, I.D.: Boundary integral equations of minimum order for the calculation of three-dimensional eddy current problems. IEEE Trans. Magn. **18**(2), 536–539 (1982)
16. Yee, K.S.: Numerical solution of initial boundary value problems involving Maxwell's equations. IEEE Trans. Antennas Propag. **14**(3), 302–307 (1966)
17. Weiland, T.: A discretization model for the solution of Maxwell's equations for six-component fields. Archiv Elektronik und Übertragungstechnik **31**, 116–120 (1977)
18. Nédélec, J.-C.: Mixed finite elements in $\mathbb{R}3$. Numerische Mathematik **35**(3), 315–341 (1980)
19. Bossavit, A.: Whitney forms: a class of finite elements for three-dimensional computations in electromagnetism. IEE Proc. A (Phys. Sci. Meas. Instrum. Manag. Edu. Rev.) **135**(8), 493–500 (1988)
20. Arnold, D.N., Falk, R.S., Winther, R.: Finite element exterior calculus, homological techniques, and applications. Acta Numerica **15**, 1–155 (2006)
21. Arnold, D., Falk, R., Winther, R.: Finite element exterior calculus: from hodge theory to numerical stability. Bull. Am. Math. Soc. **47**(2), 281–354 (2010)
22. Brenner, S.C., Scott, R.: The Mathematical Theory of Finite Element Methods, vol. 15. Springer (2008)
23. Hughes, T.J.R., Cottrell, J.A., Bazilevs, Y.: Isogeometric analysis: cad, finite elements, nurbs, exact geometry and mesh refinement. Comput. Methods Appl. Mech. Eng. **194**(39), 4135–4195 (2005)
24. Buffa, A., Sangalli, G., Vázquez, R.: Isogeometric analysis in electromagnetics: b-splines approximation. Comput. Methods Appl. Mech. Eng. **199**(17), 1143–1152 (2010)
25. Buffa, A., Rivas, J., Sangalli, G., Vázquez, R.: Isogeometric discrete differential forms in three dimensions. SIAM J. Numer. Anal. **49**(2), 818–844 (2011)
26. Beirão, L., da Veiga, A., Buffa, G.Sangalli, Vázquez, R.: Mathematical analysis of variational isogeometric methods. Acta Numerica **23**, 157–287 (2014)
27. Benzi, M., Golub, G.H., Liesen, J.: Numerical solution of saddle point problems. Acta Numerica **14**, 1–137 (2005)
28. Heise, B.: Analysis of a fully discrete finite element method for a nonlinear magnetic field problem. SIAM J. Numer. Anal. **31**(3), 745–759 (1994)
29. Hiptmair, R.: Finite elements in computational electromagnetism. Acta Numerica **11**, 237–339 (2002)

30. Alaoui, E.,, Ern, A., Vohralík, M.: Guaranteed and robust a posteriori error estimates and balancing discretization and linearization errors for monotone nonlinear problems. Comput. Methods Appl. Mech. Eng. **200**(37), 2782–2795 (2011)
31. Wachsmuth, G.: Differentiability of implicit functions. J. Math. Anal. Appl. **414**(1), 259–272 (2014). June
32. Anders, L., Mardal, K.-A., Wells, G.N.: Automated Solution of Differential Equations by the Finite Element Method. The FEniCS book, Springer (2012)
33. Freund, R.W., Hochbruck, M.: On the use of two qmr algorithms for solving singular systems and applications in markov chain modeling. Numer. Linear Algebra Appl. **1**(4), 403–420 (1994)
34. Kaasschieter, E.F.: Preconditioned conjugate gradients for solving singular systems. J. Comput. Appl. Math. **24**(1), 265–275 (1988)
35. Chaillou, A., Suri, M.: A posteriori estimation of the linearization error for strongly monotone nonlinear operators. J. Comput. Appl. Math. **205**(1), 72–87 (2007)
36. Prager, W., Synge, J.L.: Approximations in elasticity based on the concept of function space. Q. Appl. Math. **5**(3), 1–21 (1947)
37. Braess, D., Schöberl, J.: Equilibrated residual error estimator for edge elements. Math. Comput. **77**(262), 651–672 (2008)
38. Kikuchi, F., Saito, H.: Remarks on a posteriori error estimation for finite element solutions. J. Comput. Appl. Math. **199**(2), 329–336 (2007)
39. Ciarlet, P.G.: The Finite Element Method for Elliptic Problems. Elsevier (1978)
40. Römer, U., Schöps, S., Weiland, T.: Stochastic modeling and regularity of the nonlinear elliptic curl-curl equation. SIAM/ASA J Uncertainty Quantification (in press)
41. Nicolet, A., Delincé, F.: Implicit Runge-Kutta methods for transient magnetic field computation. IEEE Trans. Magn. **32**(3), 1405–1408 (1996)
42. Bartel, A., Baumanns, S., Schöps, S.: Structural analysis of electrical circuits including magnetoquasistatic devices. Appl. Numer. Math. **61**(12), 1257–1270 (2011)
43. Hairer, E., Nörsett, S.P., Wanner, G.: Solving Ordinary Differential Equations I: Nonstiff Problems. Springer Series in Computational Mathematics, 2nd edn. Springer, Berlin (2000)
44. Hairer, E., Nörsett, S.P., Wanner, G.: Solving Ordinary Differential Equations II: Stiff and Differential-Algebraic Problems. Springer Series in Computational Mathematics, 2nd edn. Springer, Berlin (2002)
45. Stiemer, M.: A Galerkin method for mixed parabolic-elliptic partial differential equations. Numerische Mathematik **116**(3), 435–462 (2010)

Chapter 4
Parametric Model, Continuity and First Order Sensitivity Analysis

The subject of this chapter is a detailed description of a parametric magnetoquasistatic model, generalizing the deterministic setting of Chap. 3. To this end we choose a continuous setting, i.e., parametrization is discussed on the differential equation level. Moreover, in a first step we allow for a general, possibly infinite dimensional parametrization, before discussing its finite dimensional approximation later on. Continuity and differentiability results will be established for different kind of inputs. In particular the sensitivity analysis presented in Sects. 4.4, 4.5 will be a key tool for propagating uncertainties in Chaps. 5 and 6.

4.1 Abstract Mathematical Reformulation of the Model

Following [1], continuity and differentiability results are conveniently discussed with the aid of a more abstract mathematical setting. Let $\boldsymbol{\beta}$ denote a generic input parameter such as the shape, the material coefficient or the imposed current excitation. Modeling of these inputs will be addressed in Sect. 4.2. Depending on the model, physical constraints on the inputs have to be imposed, e.g., the function describing the nonlinear material law has to be monotonic or the boundary of the domain has to fulfill regularity requirements. This is accounted for by introducing a set of admissibility U_{adm}. The model is rewritten implicitly by means of an operator $\mathcal{Z} : U_{\text{adm}} \times W(D) \rightarrow W^*(D)$, as

$$\mathcal{Z}(\boldsymbol{\beta}, \mathbf{A}) = 0, \tag{4.1}$$

referred to as the abstract state equation. In this context, the solution $\mathbf{A}$ to the differential equation is sometimes called the state variable. It is understood, that for each $\boldsymbol{\beta} \in U_{\text{adm}}$, we have a unique solution $\mathbf{A}$, which is a function of $\boldsymbol{\beta}$, i.e., $\mathbf{A}[\boldsymbol{\beta}] \in W(D)$. To lighten notation we also introduce $\mathbf{A}_{\boldsymbol{\beta}} := \mathbf{A}[\boldsymbol{\beta}]$, and drop the explicit dependence

U. Römer, *Numerical Approximation of the Magnetoquasistatic Model with Uncertainties*, Springer Theses, DOI 10.1007/978-3-319-41294-8_4

on $\boldsymbol{\beta}$, when no confusion is possible. Depending on the context, $\mathcal{Z}$ refers to the weak formulation of (2.24), (3.15) or the corresponding two-dimensional formulations. A Quantity of Interest (QoI) is generally denoted $Q : W(D) \to \mathbb{R}$ and assumed to be a linear functional of the solution. This is justified by observing, that QoIs in our context typically model local evaluations of fields that can be written as

$$Q(\mathbf{A}) := \int_{I_T} \int_{D_{\text{obs}}} q_1(\mathbf{A}) + q_2(\mathbf{curl}\,\mathbf{A})\, \mathrm{d}\mathbf{x}\, \mathrm{d}t, \tag{4.2}$$

with observer region $D_{\text{obs}} \subset D_{\text{E}}$ and linear functionals q_1 and q_2. To study continuity and differentiability a perturbation $\tilde{\boldsymbol{\beta}} \in \tilde{U}$ is introduced, where $\tilde{U}$ is a Banach space. Then the parameter can often be written as $\boldsymbol{\beta} = \boldsymbol{\beta}_0 + \tilde{\boldsymbol{\beta}}$, with $\boldsymbol{\beta}_0$ denoting a nominal value and we study the mappings

$$\tilde{\boldsymbol{\beta}} \mapsto \mathbf{A}[\boldsymbol{\beta}_0 + \tilde{\boldsymbol{\beta}}], \tag{4.3a}$$

$$\tilde{\boldsymbol{\beta}} \mapsto \hat{Q}[\boldsymbol{\beta}_0 + \tilde{\boldsymbol{\beta}}], \tag{4.3b}$$

for $\|\tilde{\boldsymbol{\beta}}\|_{\tilde{U}} \to 0$. In (4.3b) setting $Q(\mathbf{A}[\boldsymbol{\beta}]) =: \hat{Q}[\boldsymbol{\beta}]$ is justified, as we have already shown that (2.24) and (3.15) are uniquely solvable. Equivalently, with a smallness parameter $s > 0$, and $\boldsymbol{\beta}_s = \boldsymbol{\beta}_0 + s\tilde{\boldsymbol{\beta}}$, we set $\mathbf{A}_s := \mathbf{A}[\boldsymbol{\beta}_s]$ and $\hat{Q}_s := Q(\mathbf{A}_s)$, respectively and investigate the limit $s \to 0$. Unfortunately, in the case of shape perturbations, we do not have a simple vector space structure for U_{adm} and modeling of the perturbation is more involved.

4.2 Definition of the Model Inputs and Parametrization

Here, we identify the following three types of model inputs, important for magnetoquasistatic simulations and in particular for the simulation of accelerator magnets.

- The material coefficient, i.e., the magnetic reluctivity. It is obtained from the function f_{HB}, describing the nonlinear $H - B$ curve, which is fitted from measured data. This data is always subject to uncertainty with rather large margins in many situations. It has been observed that for some types of electrical machines, this is the most influential source of uncertainty [2].
- By definition, for iron-dominated magnets, the shape of the iron yoke is crucial for the field accuracy in the magnetic aperture. We will therefore consider the shape of the interface between the iron parts and the surrounding air as input parameter. As opposed to many applications where the boundary of the computational domain is parametrized, here, the boundary is considered to remain unchanged.

- Again, by definition, coil-dominated magnets[1] are sensitive to the coil shape as well as the current distribution and magnitude. This can be taken into account by considering the imposed current excitation as an input parameter.

The corresponding sets of admissibility have been partially described in Chap. 2 as we have formulated assumptions to guarantee the existence of a solution for $\mathcal{Z}$. In the following this will be made more precise. We fix the time t, i.e., we assume that variability will not change with respect to t. This is a reasonable assumption in many cases, but of course not in general. With a description of the parameter $\boldsymbol{\beta}$ for each specific case at hand, we will give examples of finite dimensional representations, i.e., identify vectors $\mathbf{y} \in \Gamma \subset \mathbb{R}^M$ such that $\boldsymbol{\beta} \approx \boldsymbol{\beta}_M(\mathbf{y})$. In the following $k \in \mathbb{N}_0$ will denote a regularity parameter.

4.2.1 The Material Coefficient as a Model Input

To our knowledge, considering the magnetic reluctivity as an input parameter on the equation level has not been addressed before. Key properties have been formulated in Assumption 2.2 of Chap. 2. Assuming still that the magnetic properties are piecewise constant, the aim here is to introduce variability for the iron-conductor reluctivity. Let $\mathcal{C}^k(\overline{\mathbb{R}^+})$ be the space of functions on $\mathbb{R}^+$, with bounded and uniformly continuous derivatives up to order k, endowed with the norm

$$\|u\|_{\mathcal{C}^k(\overline{\mathbb{R}^+})} := \max_{0 \leq i \leq k} \sup_{s \in \mathbb{R}^+} |\partial_s^i u(s)|, \tag{4.4}$$

see [3, p. 10]. We set $\beta_\nu := \nu_\mathrm{C}$ and define, motivated by (2.16), the set of admissibility as

$$U^\nu_{\mathrm{adm}} := \{\beta_\nu \in \mathcal{C}(\overline{\mathbb{R}^+}) \mid \beta_\nu \in [\nu_{\min}, \nu_0],\ \lim_{s\to\infty} \beta_\nu(s) = \nu_0,\ \beta_\nu(\cdot)\cdot \text{ strongly monotone and Lipschitz continuous with constants } \nu_{\min}, \nu_0\}. \tag{4.5}$$

It should be emphasized that the constants in (4.5) hold uniformly for all parameters. This implies that $\mathcal{M}$ is uniformly strongly monotone as well as uniformly Lipschitz continuous and hence, a unique solution is guaranteed for all $\beta_\nu \in U^\nu_{\mathrm{adm}}$. Let $\mathcal{C}^k_0(\overline{\mathbb{R}^+})$ denote the subspace of functions in $\mathcal{C}^k(\overline{\mathbb{R}^+})$ with compact support. We introduce a perturbed reluctivity $\beta_\nu := \beta_{\nu,0} + \tilde{\beta}_\nu$, with $\tilde{\beta}_\nu \in \mathcal{C}^1_0(\overline{\mathbb{R}^+})$ such that $\beta_\nu \in U^\nu_{\mathrm{adm}}$.

Now if the assumption on ν to be piecewise constant is dropped, the constants in (4.5) will depend on $\mathbf{x}$ and we additionally require that they hold uniformly with respect to $\mathbf{x}$. Finally, not only the reluctivity, but also the conductivity can be modeled as an input parameter. In this case, as the main requirement is the positivity,

[1] A more detailed description of coil- and iron-dominated magnets will be given in Chap. 6.

perturbations can be defined by $\beta_\sigma := \beta_{\sigma,0} + \tilde{\beta}_\sigma$, where $\beta_{\sigma,0} = \sigma_C$ and a possible definition of an admissible set would be

$$U^\sigma_{\text{adm}} := \{\beta_\sigma \in L^\infty(D_C) \mid \beta_\sigma \geq C > 0, \text{ a.e. in } D_C\}. \quad (4.6)$$

4.2.2 Approximation of the Magnetic Material Coefficient

We are going to discuss discrete representations of $H = f_{HB}(B)$, where $B := |\mathbf{B}|$ and $H := |\mathbf{H}|$. Discrete stochastic representations have been discussed in [4, 5] and also in [6]. We will partially make use of these results here. Recall that f_{HB} is subject to the constraints (2.14), in particular monotonicity is required. Although the coefficient ν is the input for the equations, measurements are typically given for f_{HB}, which will be the starting point for our discussion. The discrete coefficient is then again defined by means of the relation $\nu(\mathbf{x}, B) = f_{HB}(B)/B$ for all $\mathbf{x} \in D_C$ and $B \neq 0$.

Commonly in practice f_{HB} has to be estimated from a given table of measured tuples

$$\{\{B_i^{\text{ms}}, H_i^{\text{ms}}\}, i = 1, ..., N^{\text{ms}}, \ B_i < B_j, \ \text{for } i < j\}. \quad (4.7)$$

Many approaches presented in the literature are based on special closed form representations, often motivated directly from physical considerations, see [7, 8] and also [2] and the references therein. As an example, consider the relation

$$f_{HB}(B) = \left(c_4 + \frac{c_3 B^{c_2}}{c_1^{c_2} + B^{c_2}}\right) B, \quad (4.8)$$

used, e.g., in [9], where the parameters $\mathbf{y} = (c_1, \ldots, c_4)$ are typically determined from (4.7) by a least-squares fit. Although closed form representations as (4.8), solely based on a few parameters, are rather simple and the constraints (2.14) can be incorporated directly, the flexibility and accuracy in approximating arbitrary functions f_{HB} can be limited. As described, e.g., in [8] the Brillouin model, is not capable of accurately representing the Rayleigh region, important for applications with low magnetic field magnitudes. In contrary, spline models are a flexible tool with powerful approximation capabilities. However, besides loosing a physically motivated representation, a major difficulty is to achieve monotonicity. For monotone data (4.7), i.e., $H_i \leq H_j$ for $i < j$, the cubic spline interpolation procedure, given in [10] can be employed: on each interval $[B_i, B_{i+1}]$, f_{HB} is represented as a cubic polynomial

$$f_{HB}(B) = H_i^{\text{ms}} \phi_1(B) + H_{i+1}^{\text{ms}} \phi_2(B) + d_i \phi_3(B) + d_{i+1} \phi_4(B). \quad (4.9)$$

Here, the ϕ_i, $i = 1, \ldots, 4$, refer to the cubic Hermite basis functions and the vector $\mathbf{d} = (d_i)_{i=1}^{N^{\text{ms}}}$ approximates the derivative of f_{HB} at the end of the intervals in such a way that monotonicity is preserved. Extrapolation beyond the data range can be achieved as outlined in [11, 12]. We note that if the reluctivity is monotonic, too, a similar algorithm presented in [11] can be applied directly to ν. Given (4.9), a parametrization of the model can be achieved, by keeping $\mathbf{d}$ fixed, and setting $\mathbf{y} = \mathbf{H}^{\text{ms}}$ as we have outlined in [4]. A more general approach in a stochastic context is given in [13] and will be discussed in Chap. 5. Other spline based models have been proposed in the literature, with emphasis on preserving monotonicity by smoothing noisy data [12, 14]. However, here, as we are interested in reproducing uncertainties interpolation based techniques are favored.

4.2.3 The Shape of the Interface as a Model Input

Among the inputs under consideration here, a parametrization of the interface can be considered to be the most involved. This is due to the fact, that no vector space structure is present at first sight. Solutions to this problem have been given in the literature and the modeling of shape deformations can be considered as well-established in the context of shape optimization. In particular the velocity method [15], adapted here, is known to be a general and powerful tool. In this section, the geometrical setup given in Fig. 2.2 will be slightly modified, see Fig. 4.1. Deformations are restricted to the interior of D_{HA}, referred to as hold-all in the literature [15]. We assume, without loss of generality, that its boundary is smooth. In this section, we also assume that D_{C} is simply connected with a $\mathcal{C}^{k,1}$-regular boundary, resp. interface Γ_{I}, subject to deformations. We denote with $\mathcal{C}^{k,1}(\overline{V})$ the subspace of functions in $\mathcal{C}^{k}(\overline{V})$, that have Lipschitz continuous derivatives of order k, endowed with the Hölder norm [3, p. 10]. To assure that the coil parts always encompass the deformed iron parts $D_{\text{C},s}$ we require that $\overline{D_{\text{HA}}} \cap \overline{D_{\text{J}}} = \emptyset$. Let us recall that if Γ_{I} is of class $\mathcal{C}^{k,1}$ and connected, these properties are preserved by a transformation with a $\mathcal{C}^{k,1}$-regular, bijective map $\mathcal{T}$ [16, p. 45]. The set of admissible interfaces is defined as

$$U^{\text{I}}_{\text{adm}} := \{S \subset \mathbb{R}^3 \mid S = \mathcal{T}(\Gamma_{\text{I}}) \text{ is of class } \mathcal{C}^{0,1} \text{ and connected}, \; S \cap \partial D_{\text{HA}} = \emptyset\}. \tag{4.10}$$

As already mentioned (4.10) does not have a vector space structure and tools of differentiation in normed spaces cannot be applied. A solution is given in the context of the velocity method, described in the following. The deformation is modeled by means of a mapping $\mathcal{T}_s : \overline{D_{\text{HA}}} \to \overline{D_{\text{HA}}}$, given by $\mathcal{T}_s[\mathcal{V}](\mathbf{X}) = \mathbf{x}(s, \mathbf{X})$, where $\mathcal{V} : \overline{D_{\text{HA}}} \to \mathbb{R}^3$ and $\mathbf{x}(s) := \mathbf{x}(s, \cdot)$ [2] is the solution of the differential equation

$$\frac{\mathrm{d}\mathbf{x}}{\mathrm{d}s}(s, \mathbf{X}) = \mathcal{V}(\mathbf{x}(s, \mathbf{X})), \tag{4.11a}$$

$$\mathbf{x}(0, \mathbf{X}) = \mathbf{X}, \tag{4.11b}$$

[2] In mechanics $\mathbf{x}$ is referred to as Eulerian variable, whereas $\mathbf{X}$ denotes the Lagrangian variable.

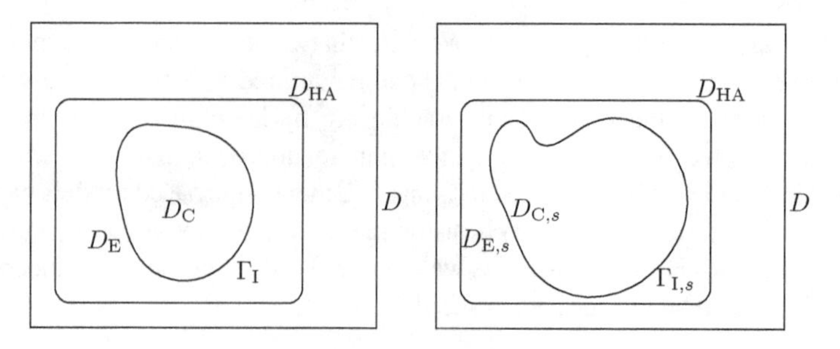

Fig. 4.1 General model geometry for parametric interfaces

see [15, p. 290]. We sometimes write $\mathcal{T}_s = \mathcal{T}_s[\boldsymbol{\mathcal{V}}]$, for short. In (4.11) $\boldsymbol{\mathcal{V}}$ denotes a velocity field, representing the direction of domain perturbations, i.e., we identify $\tilde{\boldsymbol{\beta}}_{\mathrm{I}} := \boldsymbol{\mathcal{V}}$. It is always assumed that $\tilde{\boldsymbol{\beta}}_{\mathrm{I}}$ satisfies

$$\tilde{\boldsymbol{\beta}}_{\mathrm{I}} \in \mathcal{C}^{k,1}(\overline{D_{\mathrm{HA}}})^3, \tag{4.12a}$$

$$\|\tilde{\boldsymbol{\beta}}_{\mathrm{I}}\|_{\mathcal{C}^{k,1}(\overline{D_{\mathrm{HA}}})^3} \leq 1, \tag{4.12b}$$

$$\tilde{\boldsymbol{\beta}}_{\mathrm{I}} \cdot \mathbf{n}|_{\partial D_{\mathrm{HA}}} = 0. \tag{4.12c}$$

The uniform norm constraint is needed to prevent a degeneration of the interface [17]. We obtain a family of transformed interfaces

$$\beta_{\mathrm{I,s}} := \Gamma_{\mathrm{I},s} = \mathcal{T}_s(\Gamma_{\mathrm{I}}), \tag{4.13}$$

and domains $D_{\mathrm{C},s} := \mathcal{T}_s(D_{\mathrm{C}})$, resp. $D_{\mathrm{E},s} := \mathcal{T}_s(D_{\mathrm{E}})$. There exists s_0 such that for all $s \leq s_0$ the transformed interfaces belong to $U^{\mathrm{I}}_{\mathrm{adm}}$ [16, p. 45]. Now the differentiability and continuity with respect to β_{I} ($\mathcal{V}$) can be studied in the Banach space $\mathcal{C}^{k,1}(\overline{D_{\mathrm{HA}}})^3$ (at 0). Indeed, it has been shown that this is equivalent to studying the continuity directly in $U^{\mathrm{I}}_{\mathrm{adm}}$ (slightly modified, with a suitable metric) and we refer to [15, pp. 321] for the details.

Remark 4.1 A perturbation of the coefficients is associated to a deformation of the interface Γ_{I}. Indeed, for domains parametrized as outlined above, the reluctivity and the conductivity read as

$$\nu_s(\mathbf{x}, \cdot) = \nu_0 \chi_{D_{\mathrm{E},s}}(\mathbf{x}) + \nu_{\mathrm{C}}(\cdot)\chi_{D_{\mathrm{C},s}}(\mathbf{x}), \tag{4.14}$$

$$\sigma_s(\mathbf{x}) = \sigma_{\mathrm{C}} \chi_{D_{\mathrm{C},s}}(\mathbf{x}), \tag{4.15}$$

respectively, where χ_V denotes the characteristic function of the volume V.

Note that if we generalize (4.11) to non-autonomous velocity fields $\boldsymbol{\mathcal{V}}(s, \mathbf{x}(s))$, the velocity method comprises the frequently used perturbation of the identity mapping $\mathcal{T}_s[\boldsymbol{\mathcal{V}}](\mathbf{X}) = \mathbf{X} + s\boldsymbol{\mathcal{V}}(\mathbf{X})$, see [18]. It can also be extended to rather general classes of non-smooth domains [19]. As implied by the structure theorem, that will be discussed in Sect. 4.4.2 the support of the derivative is fully contained in the interface Γ_I, hence for the actual computation only an interface velocity field is required. This gives rise to a third popular representation scheme for shape deformations, i.e., by means of normal perturbations with respect to the interface [20, 21]. Our way of computing sensitivities, as well as the discrete domain deformation technique based on splines, presented in the next section, relies on the velocity method with autonomous velocity field.

4.2.4 Approximation and Representation of Shapes

Many objects in a Computer Aided Design (CAD) environment are constructed in terms of their boundaries [22]. If the boundary shape has a simple form, which is the case, e.g., for bricks, cylinders or spheres, parametrization can easily be done. As an example, Fig. 4.2 depicts a variable hyperbolic shape profile, determined by a single parameter. Perturbing only this single parameter of the hyperbola lacks flexibility as all perturbed shapes will be again of the same kind. A more general representation is given by means of NURBS. Based on the observation that most of the shapes used in CAD are conic sections and NURBS are capable of representing them exactly, this is often called an exact geometry representation [23]. In local coordinates $\xi, \eta \in (0, 1)$, a line and a surface are given by

$$\mathbf{F}(\xi) = \sum_{i=1}^{N_{\text{cp}}} R_i^q(\xi)\mathbf{c}_i, \tag{4.16}$$

$$\mathbf{F}(\xi, \eta) = \sum_{i=1}^{N_{\text{cp}}} \sum_{j=1}^{N_{\text{cp}}} R_i^q(\xi) R_j^q(\eta)\mathbf{c}_{i,j}, \tag{4.17}$$

respectively. Here, R_i^q and R_j^q denote NURBS basis functions of degree q and regularity C^k, $k \geq 0$, defined as

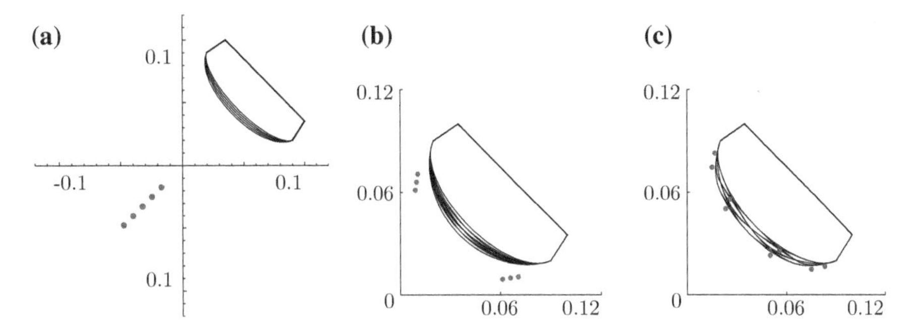

Fig. 4.2 Uncertain shape. The control points, marked in *red*, are perturbed uniformly in an interval with 80 % deviation from the nominal value for **a** and 20 % deviation for the remaining refinement levels, respectively. **a** One control point. **b** Two control points. **c** Four control points

$$R_i^q(\xi) = \frac{B_i^q(\xi) w_i}{\sum_{i=1}^{N_{\text{cp}}} B_i^q(\xi) w_i} \tag{4.18}$$

with B-spline basis function B_i^q and weights w_i. We also refer to [22, 23] for details and Appendix B for more on B-splines. Here, the parameters can be chosen as $\mathbf{y} = (\mathbf{c}_{i,j})_{i,j=1}^{N_{\text{cp}}}$. A volumetric model, as employed in Sect. 3.3.2 and required for the velocity method, is obtained from (4.17) by a completion procedure [22], e.g., the spring model [24] which is adapted here. Perturbing the control points as $\mathbf{c}_{i,j}(s) = (\mathbf{c}_0)_{i,j} + s\tilde{\mathbf{c}}_{i,j}$ results in a surface

$$\mathbf{F}_s(\xi, \eta) = \sum_{i=1}^{N_{\text{cp}}} \sum_{j=1}^{N_{\text{cp}}} R_i^q(\xi) R_j^q(\eta) \left((\mathbf{c}_0)_{i,j} + s\tilde{\mathbf{c}}_{i,j} \right). \tag{4.19}$$

In particular we identify $\mathbf{X} = \mathbf{F}_0$ and $\mathbf{x}(s) = \mathbf{F}_s \circ \mathbf{F}_0^{-1}$ in (4.11). Then computing the velocity field as

$$\mathcal{V}(\mathbf{x}) = \frac{\partial \mathcal{T}_s}{\partial s} \circ \mathcal{T}_s^{-1}(\mathbf{x}), \tag{4.20}$$

we obtain

$$\mathcal{V} \circ \mathbf{F}_s = \sum_{i,j} R_i^q(\xi) R_j^q(\eta) \tilde{\mathbf{c}}_{i,j}. \tag{4.21}$$

Note that the weights w_i could also be perturbed, however, this would result in more complicated expressions for the velocity field. We also refer to [25] for an application in mechanics and to [26] for a related approach to sensitivity analysis in the context of isogeometric analysis. Now, considering again Fig. 4.2, the hyperbolas are represented by means of a NURBS curve with a different number of control points. In red are depicted those control points that are subject to variation, i.e., one, two and four points, respectively. With one control point solely, Fig. 4.2a, nothing is gained with respect to taking the parameter of the hyperbola. However, as can be seen in Fig. 4.2b, c, NURBS allow for a more flexible modeling of shape perturbations. Note, however, that the perturbation magnitude of the control points is not related to the real perturbation of the shape anymore: in Fig. 4.2a, 80 % variability in the control point results in a deviation of 0.015 in the center of the curve, whereas for 20 % variability in Fig. 4.2b, c, the curve is perturbed by 0.023.

In the iso-parametric finite element context, shapes are approximated by, possibly curved, polyhedrals. The geometry mapping in this case can be written locally for an element of the mesh as

$$\mathbf{F}(\xi, \eta) = \sum_i N_i(\xi, \eta) \mathbf{c}_i, \tag{4.22}$$

where N_i refers to the i-th nodal shape function. Note that only shapes of C^0 regularity can be obtained in this context.

4.2.5 *The Source Current Density as a Model Input*

Here, we simply recall that by Assumption 3.2 the coil current, expressed as the imposed current excitation has to be weak divergence free and square-integrable, i.e., we set

$$U_{\text{adm}}^{\mathbf{J}} := \mathcal{H}(\text{div } 0, D). \tag{4.23}$$

Perturbations are given by $\boldsymbol{\beta}_{\text{J}} = \boldsymbol{\beta}_{\text{J},0} + \tilde{\boldsymbol{\beta}}_{\text{J}}$ and based on the assumption

$$\tilde{\boldsymbol{\beta}}_{\text{J}} \in \mathcal{H}(\text{div } 0, D), \tag{4.24}$$

$\boldsymbol{\beta}_{\text{J}}$ is admissible.

4.2.6 *Conductor Models*

Here, we are looking for a discrete representation of the imposed source current density $\mathbf{J}$ of a conductor. In a first step, given the total current I, we set $\mathbf{J} = \boldsymbol{\chi} I$, where the vector $\boldsymbol{\chi} \in \mathcal{H}(\text{div } 0, D)$ denotes the *winding* function [27]. The distribution of $\boldsymbol{\chi}$ depends on the conductor type under consideration. In general, there is a distinction between stranded and solid conductors, respectively [28]. Solid conductors, i.e., conductors with a current displacement due to the term $\sigma \partial \mathbf{A}/\partial t$ are not considered here, as we recall that $D_{\text{J}} \subseteq D_{\text{E}}$. Instead, in the stranded conductor model [28], currents are supposed to be uniformly distributed over the conductor's cross-section. This modeling addresses filamentary conductors composed of a large number of thin strands, such that the strand diameter is small compared to the skin depth and too small to be resolved by a grid based model. The piecewise constant current in each strand cross-section is then homogenized over the whole conductor. For very simple geometries, closed form expressions can be found for the winding function. For instance in the case of a straight conductor in z-direction and strands connected in series, we have $\boldsymbol{\chi} = N_{\text{str}}/A_{\text{C}} \mathrm{e}_z$, where e_z refers to the unit vector in z-direction. Here, N_{str} denotes the number of strands and A_{C} the cross-section of the conductor. For more complex geometries an (anisotropic) electrokinetic problem

$$\text{div } (\boldsymbol{\sigma}_{\text{str}} \, \mathbf{grad} \, \phi) = 0, \quad \text{in } D_{\text{J}} \tag{4.25a}$$

$$\mathbf{grad} \, \phi \cdot \mathbf{n} = 0, \quad \text{on } \partial D_{\text{J}}, \tag{4.25b}$$

$$\int_{\Sigma} \mathbf{grad} \, \phi \cdot \mathbf{n} \, \mathrm{d}\mathbf{x} = 1, \tag{4.25c}$$

with conductivity tensor $\boldsymbol{\sigma}_{\text{str}}$, for a cut Σ of D_{J}, can be solved by the finite element method. Then, after homogenization of the current in the cross-sections if necessary, as described, e.g., in [29], one obtains for the stranded conductor $\boldsymbol{\chi} = N_{\text{str}} \mathbf{grad} \, \phi$. The current constraint (4.25c) can be imposed by a Lagrange multiplier technique [30].

A further aim of conductor models is to allow for a circuit coupling of the field model. Indeed, as the equations then comprise the total current, as shown above, they can be supplemented by the circuit equation

$$\frac{\mathrm{d}}{\mathrm{d}t}\int_D \chi \cdot \mathbf{A}\,\mathrm{d}\mathbf{x} + RI = V, \tag{4.26}$$

with the circuit voltage V, the conductor's DC resistance R and the integral term representing the flux linkage [27]. Note that, although the conductivity has been neglected with regard to the current displacement, $\boldsymbol{\sigma}_{\mathrm{str}}$ nevertheless gives rise to the conductor resistance. For details on the coupling of the magnetoquasistatic model to networks we refer to [31].

Depending on the context the model parameter can be the current or the voltage, i.e., $y = \{I, V\}$. Other conductor modeling techniques, such as the Roebelbar, where strands are transposed to minimize losses, or the foil conductor, see, e.g., [27] and the references therein, are not considered here. Note that in a more general context we could chose not only I, V, but any other coil property, such as the shape of the boundary of $D_{\mathbf{J}}$, or the number of windings as a parameter.

4.3 Continuity with Respect to the Input Data

As we have argued, the continuity with respect to the input data is a crucial property of a mathematical model. For this investigation we consider the static case, solely, and refer to [32] for continuity results for the time harmonic, linear magnetoquasistatic model. Consider the compound input vector, defined as follows: every point $\boldsymbol{\beta}_0 = (\nu_{\mathrm{C}}, \Gamma_{\mathrm{I}}, \mathbf{J}_0)$ is perturbed by means of $\tilde{\boldsymbol{\beta}} \in \mathcal{C}_0^1(\overline{\mathbb{R}^+}) \times \mathcal{C}^{0,1}(\overline{D_{\mathrm{HA}}})^3 \times \mathcal{H}(\mathrm{div}\,0, D) = \tilde{U}$ as outlined in the previous sections. Then we can find an s_0, such that for all $\mathbf{s} = (s_1, s_2, s_3) \le s_0$ (understood component-wise)

$$\boldsymbol{\beta}_{\mathbf{s}} \in U^{\nu}_{\mathrm{adm}} \times U^{\mathrm{I}}_{\mathrm{adm}} \times U^{\mathbf{J}}_{\mathrm{adm}} =: U_{\mathrm{adm}}. \tag{4.27}$$

The parametrized version of (3.14) reads, find $\mathbf{A}_{\mathbf{s}} \in W_{\mathrm{st}}(D)$ such that

$$(\beta_{\nu,s_1}(|\mathbf{curl}\,\mathbf{A}_{\mathbf{s}}|)\mathbf{curl}\,\mathbf{A}_{\mathbf{s}}, \mathbf{curl}\,\mathbf{v})_{D_{\mathrm{C},s_2}} + (\nu_0\mathbf{curl}\,\mathbf{A}_{\mathbf{s}}, \mathbf{curl}\,\mathbf{v})_{D_{\mathrm{E},s_2}} = (\boldsymbol{\beta}_{\mathbf{J},s_3}, \mathbf{v})_D, \tag{4.28}$$

for all $\mathbf{v} \in W_{\mathrm{st}}(D)$. We observe, that by the definition of U_{adm}, $\mathbf{A}_{\mathbf{s}}$ exists for all $\boldsymbol{\beta}_{\mathbf{s}} \in U_{\mathrm{adm}}$. Continuity is now stated and proven in the following result.

Proposition 4.1 *Let* $\mathbf{A}$ *be the solution of (4.28) with* $\boldsymbol{\beta}_0$. *For* $\boldsymbol{\beta}_{\mathbf{s}} \in U_{\mathrm{adm}}$ *given by (4.27), the solution* $\mathbf{A}_{\mathbf{s}}$ *of (4.28) converges to* $\mathbf{A}_0 = \mathbf{A}$ *as* $\mathbf{s} \to 0$.

Proof To simplify notation we define the operator $\mathcal{M}_{s_1,s_2}$ by

$$< \mathcal{M}_{s_1,s_2}\mathbf{u}, \mathbf{v} >:= (\beta_{\nu,s_1}(|\mathbf{curl}\,\mathbf{u}|)\mathbf{curl}\,\mathbf{u}, \mathbf{curl}\,\mathbf{v})_{D_{\mathrm{C},s_2}} + (\nu_0\mathbf{curl}\,\mathbf{u}, \mathbf{curl}\,\mathbf{v})_{D_{\mathrm{E},s_2}}, \tag{4.29}$$

for $\mathbf{u}, \mathbf{v} \in W_{\mathrm{st}}(D)$. Equation (4.28) is rewritten as

$$< \mathcal{M}_{s_1,s_2}\mathbf{A_s}, \mathbf{v} > - < \mathcal{M}_{s_1,s_2}\mathbf{A}, \mathbf{v} >= (\boldsymbol{\beta}_{\mathbf{J},s_3}, \mathbf{v})_D - < \mathcal{M}_{s_1,s_2}\mathbf{A}, \mathbf{v} > . \tag{4.30}$$

Observing that $\mathcal{M}_{s_1,s_2}$ is strongly monotone, for $\mathbf{v} = \mathbf{A_s} - \mathbf{A}$ we obtain

$$\nu_{\min}\|\mathbf{A_s} - \mathbf{A}\|^2_{W_{\mathrm{st}}(D)} \leq (\boldsymbol{\beta}_{\mathbf{J},s_3}, \mathbf{A_s} - \mathbf{A})_D - < \mathcal{M}_{s_1,s_2}\mathbf{A}, \mathbf{A_s} - \mathbf{A} > . \tag{4.31}$$

Adding and subtracting terms, using the state equation (4.28) for $\mathbf{s} = 0$ and the triangle inequality yields

$$\begin{aligned}\nu_{\min}\|\mathbf{A_s} - \mathbf{A}\|^2_{W_{\mathrm{st}}(D)} \leq & \,| < \mathcal{M}_{s_1,0}\mathbf{A}, \mathbf{A_s} - \mathbf{A} > - < \mathcal{M}_{s_1,s_2}\mathbf{A}, \mathbf{A_s} - \mathbf{A} > | \\ & + | < \mathcal{M}\mathbf{A}, \mathbf{A_s} - \mathbf{A} > - < \mathcal{M}_{s_1,0}\mathbf{A}, \mathbf{A_s} - \mathbf{A} > | \\ & + |(\boldsymbol{\beta}_{\mathbf{J},s_3}, \mathbf{A_s} - \mathbf{A})_D - (\boldsymbol{\beta}_{\mathbf{J},0}, \mathbf{A_s} - \mathbf{A})_D|.\end{aligned}$$

The definition of the input perturbations and a change of variables

$$\begin{aligned}< \mathcal{M}_{s_1,0}\mathbf{A} - \mathcal{M}_{s_1,s_2}\mathbf{A}, \mathbf{u} >= & \int_{D_{\mathrm{C}}} (1 - \det(D\mathcal{T}_{s_2}))\beta_{\nu,s_1}(|\mathbf{curl}\,\mathbf{A}|)\mathbf{curl}\,\mathbf{A} \cdot \mathbf{curl}\,\mathbf{u}\,\mathrm{d}\mathbf{x} \\ & + \int_{D_{\mathrm{E}}} (1 - \det(D\mathcal{T}_{s_2}))\nu_0\mathbf{curl}\,\mathbf{A} \cdot \mathbf{curl}\,\mathbf{u}\,\mathrm{d}\mathbf{x} \\ = & \int_D (1 - \det(D\mathcal{T}_{s_2}))\nu_{s_1}(\cdot, |\mathbf{curl}\,\mathbf{A}|)\mathbf{curl}\,\mathbf{A} \cdot \mathbf{curl}\,\mathbf{u}\,\mathrm{d}\mathbf{x},\end{aligned}$$

for s_2 small [15, p. 342], in turn yield

$$\begin{aligned}\nu_{\min}\|\mathbf{A_s} - \mathbf{A}\|^2_{W_{\mathrm{st}}(D)} \leq & \, s_1|(\tilde{\beta}_\nu(|\mathbf{curl}\,\mathbf{A}|)\mathbf{curl}\,\mathbf{A}, \mathbf{curl}\,(\mathbf{A_s} - \mathbf{A}))_{D_{\mathrm{C},s_2}}| \\ & + |(|1 - \det(D\mathcal{T}_{s_2}))|\nu_{s_1}(\cdot, |\mathbf{curl}\,\mathbf{A}|)\mathbf{curl}\,\mathbf{A}, \mathbf{curl}\,(\mathbf{A_s} - \mathbf{A}))_D| \\ & + s_3|(\tilde{\boldsymbol{\beta}}_\mathbf{J}, \mathbf{A_s} - \mathbf{A})_D|.\end{aligned}$$

We apply the Cauchy-Schwarz and the Poincaré-Friedrich's inequality (3.12) and obtain

$$\|\mathbf{A_s} - \mathbf{A}\|_{W_{\mathrm{st}}(D)} \leq \left(s_1\|\tilde{\beta}_\nu\|_{C(\overline{\mathbb{R}_0^+})} + \nu_0\|1 - \det(D\mathcal{T}_{s_2})\|_{L^\infty(D)} + s_3 C_{\mathrm{F}}\|\tilde{\boldsymbol{\beta}}_\mathbf{J}\|_2\right) \frac{\|\mathbf{curl}\,\mathbf{A}\|_{W_{\mathrm{st}}(D)}}{\nu_{\min}}. \tag{4.32}$$

Finally, passing to the limit $\mathbf{s} \to 0$ yields the result as $\lim_{s_2\to 0}\det(D\mathcal{T}_{s_2}) = 1$. □

We conclude by observing, that this result also implies the continuity of $\hat{Q}$ w.r.t. $\mathbf{s}$ due to the linearity of the QoI.

4.4 Sensitivity Analysis, Direct Approach

For sensitivity analysis our focus remains on the static formulation for simplicity. An extension to the time transient case will be discussed in Sect. 4.6. Several contributions should be mentioned concerning the sensitivity analysis of magnetoquasistatic/magnetostatic equations. In particular, for shape sensitivity analysis we refer to [9, 33–37]. Similar results are obtained here, however in the context of the velocity method. Additionally, direct methods will be discussed in more detail. They are in particular appealing for low-dimensional representations of input parameters. Proposition 4.3 is based on [38], where a linear elliptic interface problem has been considered, but apparently new in the present context. Moreover, results with respect to the nonlinear reluctivity as we have partially published in [4] have not been stated elsewhere, to our knowledge. Not related to magnetoquasistatics, but of general importance for this section are [1, 21, 38].

For the static case, the state equation is given in strong form as

$$
\begin{aligned}
\mathbf{curl}\,(\nu(\cdot, |\mathbf{curl}\,\mathbf{A}|)\mathbf{curl}\,\mathbf{A}) &= \mathbf{J}, && \text{in } D_{\mathrm{C}} \cup D_{\mathrm{E}}, && (4.33a)\\
[\![\mathbf{A}]\!]_{\Gamma_{\mathrm{I}}} &= 0, && \text{on } \Gamma_{\mathrm{I}}, && (4.33b)\\
[\![\nu(\cdot, |\mathbf{curl}\,\mathbf{A}|)\mathbf{curl}\,\mathbf{A}]\!]_{\Gamma_{\mathrm{I}}} &= 0, && \text{on } \Gamma_{\mathrm{I}}, && (4.33c)\\
\mathbf{A} \times \mathbf{n} &= 0, && \text{on } \Gamma_{\mathrm{D}}, && (4.33d)\\
\operatorname{div} \mathbf{A} &= 0, && \text{in } D, && (4.33e)
\end{aligned}
$$

where we recall that $[\![\mathbf{u}]\!]_{\Gamma_{\mathrm{I}}} := \mathbf{u}^+ \times \mathbf{n}^+ + \mathbf{u}^- \times \mathbf{n}^-$ denotes the jump operator at the interface. As outlined in Sect. 4.2, we consider perturbations of the inputs $\boldsymbol{\beta}_s$, $s > 0$. Here, the aim of computation is the derivative

$$
Q' = \lim_{s \to 0} \frac{Q_s - Q}{s}, \qquad (4.34)
$$

referred to as gradient. To this end we first derive an equation for the Gâteaux derivative $\mathbf{A}'$ (see Sect. 3.2 for a definition) of the magnetic vector potential for each input separately. In the spirit of [39], for all given inputs, $\mathbf{A}'$ can be characterized by means of a boundary value problem, which in our case reads as

$$
\begin{aligned}
\mathbf{curl}\,\left(\boldsymbol{\nu}_{\mathrm{d}}(\cdot, \mathbf{curl}\,\mathbf{A})\mathbf{curl}\,\mathbf{A}'\right) &= \mathbf{G}, && \text{in } D_{\mathrm{C}} \cup D_{\mathrm{E}}, && (4.35a)\\
[\![\mathbf{A}']\!]_{\Gamma_{\mathrm{I}}} &= \mathbf{g}_1, && \text{on } \Gamma_{\mathrm{I}}, && (4.35b)\\
[\![\boldsymbol{\nu}_{\mathrm{d}}(\cdot, \mathbf{curl}\,\mathbf{A})\mathbf{curl}\,\mathbf{A}']\!]_{\Gamma_{\mathrm{I}}} &= \mathbf{g}_2, && \text{on } \Gamma_{\mathrm{I}}, && (4.35c)\\
\mathbf{A}' \times \mathbf{n} &= 0, && \text{on } \Gamma_{\mathrm{D}}, && (4.35d)\\
\operatorname{div} \mathbf{A}' &= 0, && \text{in } D, && (4.35e)
\end{aligned}
$$

where the terms $\mathbf{G}$ and $\mathbf{g}_{1,2}$ depend on $\tilde{\boldsymbol{\beta}}$ and $\mathbf{A}$. Except for interface variations, Fréchet differentiability is also discussed.

Solving problem (4.35) directly might be beneficial when investigating multiple quantities of interest and a small number of possible input perturbations. Although, the tangential component of both the vector potential and the magnetic field can exhibit jumps at the interface, (4.35) can be approximated with optimal order of convergence by appropriate techniques, e.g., the offset technique used in [38]. With the derivative of $\mathbf{A}_s$ at hand, the gradient is expressed as

$$Q' = \int_{D_{\text{obs}}} q_1(\mathbf{A}') + q_2(\mathbf{curl}\,\mathbf{A}')\;\mathrm{d}\mathbf{x}. \tag{4.36}$$

For the purpose of sensitivity analysis in general, we assume for the remaining part of Sect. 4, that Assumption 3.10 is satisfied. We proceed establishing (4.35) for the different kind of inputs.

4.4.1 *Magnetic Material Coefficient Sensitivity*

The differentiability of the vector potential is established in the following result. Note that the assumptions on the regularity of the solution imposed in order to show Fréchet differentiability, might be too restrictive in many practical setups.

Proposition 4.2 *Let the input parameter be given by* $\beta_\nu = \beta_{\nu,0} + \tilde{\beta}_\nu$ *as described in Sect. 4.2.1. The mapping* $\tilde{\beta}_\nu \mapsto \mathbf{A}[\beta_{\nu,0} + \tilde{\beta}_\nu]$ *is Gâteaux differentiable with* $\mathbf{A}' \in W_{\text{st}}(D)$ *as the weak solution of*

$$\begin{aligned}
\mathbf{curl}\left(\boldsymbol{\nu}_{\mathrm{d}}(\cdot,\mathbf{curl}\,\mathbf{A})\mathbf{curl}\,\mathbf{A}'\right) &= -\mathbf{curl}\left(\tilde{\beta}_\nu(|\mathbf{curl}\,\mathbf{A}|)\mathbf{curl}\,\mathbf{A}\right), && \text{in } D_{\mathrm{C}}, && (4.37a)\\
\mathbf{curl}\left(\nu_0\mathbf{curl}\,\mathbf{A}'\right) &= 0, && \text{in } D_{\mathrm{E}}, && (4.37b)\\
[\![\mathbf{A}']\!]_{\Gamma_{\mathrm{I}}} &= 0, && \text{on } \Gamma_{\mathrm{I}}, && (4.37c)\\
[\![\boldsymbol{\nu}_{\mathrm{d}}(\cdot,\mathbf{curl}\,\mathbf{A})\mathbf{curl}\,\mathbf{A}']\!]_{\Gamma_{\mathrm{I}}} &= -\tilde{\beta}_\nu(|\mathbf{curl}\,\mathbf{A}|)\mathbf{curl}\,\mathbf{A}\times\mathbf{n}, && \text{on } \Gamma_{\mathrm{I}}, && (4.37d)\\
\mathbf{A}'\times\mathbf{n} &= 0, && \text{on } \Gamma_{\mathrm{D}}, && (4.37e)\\
\operatorname{div}\,\mathbf{A}' &= 0, && \text{in } D. && (4.37f)
\end{aligned}$$

If, moreover, for all $\mathbf{r},\mathbf{s}\in\mathbb{R}^3$

$$|D\mathbf{h}(\cdot,\mathbf{r}) - D\mathbf{h}(\cdot,\mathbf{s})|_{\mathbb{R}^3\times\mathbb{R}^3} \le C_{\mathbf{h}}'|\mathbf{r}-\mathbf{s}|_{\mathbb{R}^3} \tag{4.38}$$

holds and the solution $\mathbf{A}_\nu$ *associated to* β_ν *satisfies*

$$\|\mathbf{curl}\,(\mathbf{A}_\nu - \mathbf{A})\|_{L^\infty(D)^3} \to 0, \tag{4.39}$$

for $\tilde{\beta}_\nu \to 0$*, then* $\mathbf{A}'$ *can be identified with the Fréchet derivative.*

Proof The proof follows the lines of [40, Lemma 1], where a linear elliptic problem was considered, but requires some modifications due to the nonlinearity. We consider

$\beta_\nu = \beta_{\nu,0} + s\tilde{\beta}_\nu$ with $s > 0$. The perturbed and unperturbed state equation read as

$$(\nu_{\mathrm{C}}(|\mathbf{curl}\,\mathbf{A}|)\mathbf{curl}\,\mathbf{A}, \mathbf{curl}\,\mathbf{v})_{D_{\mathrm{C}}} + (\nu_0\mathbf{curl}\,\mathbf{A}, \mathbf{curl}\,\mathbf{v})_{D_{\mathrm{E}}} = (\mathbf{J}, \mathbf{v})_D, \tag{4.40a}$$

$$(\beta_{\nu,s}(|\mathbf{curl}\,\mathbf{A}_s|)\mathbf{curl}\,\mathbf{A}_s, \mathbf{curl}\,\mathbf{v})_{D_{\mathrm{C}}} + (\nu_0\mathbf{curl}\,\mathbf{A}_s, \mathbf{curl}\,\mathbf{v})_{D_{\mathrm{E}}} = (\mathbf{J}, \mathbf{v})_D, \tag{4.40b}$$

for all $\mathbf{v} \in \mathcal{C}_0^\infty(D)^3 \cap \mathcal{H}(\mathrm{div}\,0, D)$. Taking the difference of both equations, as well as adding and subtracting terms, yields for $\mathbf{A}$ and $\mathbf{A}_s \in W_{\mathrm{st}}(D)$

$$\begin{aligned}(\nu_{\mathrm{C}}(|\mathbf{curl}\,\mathbf{A}_s|)\mathbf{curl}\,\mathbf{A}_s - \nu_{\mathrm{C}}(|\mathbf{curl}\,\mathbf{A}|)\mathbf{curl}\,\mathbf{A}, \mathbf{curl}\,\mathbf{v})_{D_{\mathrm{C}}} + (\nu_0\mathbf{curl}\,(\mathbf{A}_s - \mathbf{A}), \mathbf{curl}\,\mathbf{v})_{D_{\mathrm{E}}} \\ = -((\beta_{\nu,s}(|\mathbf{curl}\,\mathbf{A}_s|) - \nu_{\mathrm{C}}(|\mathbf{curl}\,\mathbf{A}_s|))\mathbf{curl}\,\mathbf{A}_s, \mathbf{curl}\,\mathbf{v})_{D_{\mathrm{C}}}.\end{aligned} \tag{4.41}$$

Dividing by s and taking the limit $s \to 0$ we obtain by Lemma 3.9 for the Gâteaux derivative $\mathbf{A}'$,

$$\begin{aligned}(\boldsymbol{\nu}_{\mathrm{d}}(\cdot, \mathbf{curl}\,\mathbf{A})\mathbf{curl}\,\mathbf{A}', \mathbf{curl}\,\mathbf{v})_{D_{\mathrm{C}}} + (\nu_0\mathbf{curl}\,\mathbf{A}', \mathbf{curl}\,\mathbf{v})_{D_{\mathrm{E}}} = \\ -(\tilde{\beta}_\nu(|\mathbf{curl}\,\mathbf{A}|)\mathbf{curl}\,\mathbf{A}, \mathbf{curl}\,\mathbf{v})_{D_{\mathrm{C}}}.\end{aligned} \tag{4.42}$$

To obtain (4.37) we formally proceed as follows: as $\mathbf{A}' \in W_{\mathrm{st}}(D)$ (4.37e) and (4.37f) hold and the jump of the tangential trace vanishes (4.37c). We recall the integrating by parts formula

$$(\mathbf{u}, \mathbf{curl}\,\mathbf{v})_D - (\mathbf{curl}\,\mathbf{u}, \mathbf{v})_D = (\mathbf{u} \times \mathbf{n}, \mathbf{v})_{\Gamma_{\mathrm{I}}}. \tag{4.43}$$

Applying (4.43) to (4.42) yields

$$\begin{aligned}(\mathbf{curl}\,(\boldsymbol{\nu}_{\mathrm{d}}(\cdot, \mathbf{curl}\,\mathbf{A})\mathbf{curl}\,\mathbf{A}' + \tilde{\beta}_\nu(|\mathbf{curl}\,\mathbf{A}|)\mathbf{curl}\,\mathbf{A}), \mathbf{v})_{D_{\mathrm{C}}} + (\mathbf{curl}\,(\nu_0\mathbf{curl}\,\mathbf{A}'), \mathbf{v})_{D_{\mathrm{E}}} \\ + ([\![\boldsymbol{\nu}_{\mathrm{d}}(\cdot, \mathbf{curl}\,\mathbf{A})\mathbf{curl}\,\mathbf{A}']\!]_{\Gamma_{\mathrm{I}}} + \tilde{\beta}_\nu(|\mathbf{curl}\,\mathbf{A}|)\mathbf{curl}\,\mathbf{A} \times \mathbf{n}, \mathbf{v})_{\Gamma_{\mathrm{I}}} = 0,\end{aligned} \tag{4.44}$$

where we recall that $\mathbf{n}$ is the exterior unit normal of the domain D_{C}. Then, relations (4.37a), (4.37b) follow by choosing test functions $\mathbf{v} \in \mathcal{C}_0^\infty(D_{\mathrm{C}})^3$ and $\mathbf{v} \in \mathcal{C}_0^\infty(D_{\mathrm{E}})^3$, respectively, whereas choosing test functions $\mathbf{v} \in \mathcal{C}^\infty(\Gamma_{\mathrm{I}})^3$ yields (4.37d).

Concerning Fréchet differentiability, we first introduce for $\mathbf{u}, \mathbf{v}, \mathbf{w} \in W_{\mathrm{st}}(D)$

$$\begin{aligned}< \mathcal{M}_{\mathrm{L}}(\overline{\mathbf{uv}})(\mathbf{u} - \mathbf{v}), \mathbf{w} > &:= \int_D \int_0^1 \partial_t \mathbf{h}(\cdot, \mathbf{curl}\,(\mathbf{v} + t(\mathbf{u} - \mathbf{v}))) \cdot \mathbf{w}\, \mathrm{d}t\, \mathrm{d}\mathbf{x}, \\ &= \int_0^1 a'(\mathbf{v} + t(\mathbf{u} - \mathbf{v}); \mathbf{u} - \mathbf{v}, \mathbf{w})\, \mathrm{d}t,\end{aligned}$$

see [41] such that

$$\mathcal{M}(\mathbf{A}_s) - \mathcal{M}(\mathbf{A}) = \mathcal{M}_{\mathrm{L}}(\overline{\mathbf{A}_s\mathbf{A}})(\mathbf{A}_s - \mathbf{A}). \tag{4.45}$$

We observe that by Assumption 3.10 $\mathcal{M}_{\mathrm{L}}(\overline{\mathbf{A}_s\mathbf{A}})$ is coercive and hence

$$\|\mathbf{A}_s - \mathbf{A} - s\mathbf{A}'\|^2_{W_{\mathrm{st}}(D)} \leq C| < \mathcal{M}_{\mathrm{L}}(\overline{\mathbf{A}_s\mathbf{A}})(\mathbf{A}_s - \mathbf{A} - s\mathbf{A}'), \mathbf{A}_s - \mathbf{A} - s\mathbf{A}' > |. \tag{4.46}$$

Using (4.45) we infer

$$\|\mathbf{A}_s - \mathbf{A} - s\mathbf{A}'\|_{W_{\mathrm{st}}(D)} \leq C \sup_{\mathbf{v}\in W_{\mathrm{st}}(D)} \frac{| < \mathcal{M}(\mathbf{A}_s) - \mathcal{M}(\mathbf{A}) - s\mathcal{M}_{\mathrm{L}}(\overline{\mathbf{A}_s\mathbf{A}})\mathbf{A}', \mathbf{v} > |}{\|\mathbf{v}\|_{W_{\mathrm{st}}(D)}}. \tag{4.47}$$

Let

$$< \tilde{\mathcal{M}}(\mathbf{A}_s), \mathbf{v} > := \int_{D_{\mathrm{C}}} \tilde{\beta}_\nu(|\mathbf{curl}\,\mathbf{A}_s|)\mathbf{curl}\,\mathbf{A}_s \cdot \mathbf{curl}\,\mathbf{v}\,\mathrm{d}\mathbf{x}. \tag{4.48}$$

Adding $s\tilde{\mathcal{M}}(\mathbf{A}_s) - s\tilde{\mathcal{M}}(\mathbf{A}_s)$ and using (4.40) yields

$$\begin{aligned}
&\|\mathbf{A}_s - \mathbf{A} - s\mathbf{A}'\|_{W_{\mathrm{st}}(D)} \leq \\
&C \sup_{\mathbf{v}\in W_{\mathrm{st}}(D)} \frac{1}{\|\mathbf{v}\|_{W_{\mathrm{st}}(D)}} \underbrace{|(\beta_{\nu,s}(|\mathbf{curl}\,\mathbf{A}_s|)\mathbf{curl}\,\mathbf{A}_s, \mathbf{curl}\,\mathbf{v})_D - < \mathcal{M}(\mathbf{A}), \mathbf{v} > |}_{=0} \\
&+C \sup_{\mathbf{v}\in W_{\mathrm{st}}(D)} \frac{1}{\|\mathbf{v}\|_{W_{\mathrm{st}}(D)}} s| < \tilde{\mathcal{M}}(\mathbf{A}_s) + \mathcal{M}_{\mathrm{L}}(\overline{\mathbf{A}_s\mathbf{A}})\mathbf{A}', \mathbf{v} > |.
\end{aligned}$$

Adding in turn $\mathcal{M}_{\mathrm{L}}\ \mathbf{A}' - \mathcal{M}_{\mathrm{L}}\ \mathbf{A}'$, with $\mathcal{M}_{\mathrm{L}}$ linearized at $\mathbf{A}$ such that $\mathcal{M}_{\mathrm{L}}\ \mathbf{A}' = -\tilde{\mathcal{M}}(\mathbf{A})$ we obtain

$$\begin{aligned}
&\|\mathbf{A}_s - \mathbf{A} - s\mathbf{A}'\|_{W_{\mathrm{st}}(D)} \leq \\
&C \sup_{\mathbf{v}\in W_{\mathrm{st}}(D)} \frac{1}{\|\mathbf{v}\|_{W_{\mathrm{st}}(D)}} s| < \tilde{\mathcal{M}}(\mathbf{A}_s) + \mathcal{M}_{\mathrm{L}}\ \mathbf{A}' - \mathcal{M}_{\mathrm{L}}\ \mathbf{A}' + \mathcal{M}_{\mathrm{L}}(\overline{\mathbf{A}_s\mathbf{A}})\mathbf{A}', \mathbf{v} > | \\
&=C \sup_{\mathbf{v}\in W_{\mathrm{st}}(D)} \frac{1}{\|\mathbf{v}\|_{W_{\mathrm{st}}(D)}} s| < \tilde{\mathcal{M}}(\mathbf{A}_s) - \tilde{\mathcal{M}}(\mathbf{A}) - \mathcal{M}_{\mathrm{L}}\ \mathbf{A}' + \mathcal{M}_{\mathrm{L}}(\overline{\mathbf{A}_s\mathbf{A}})\mathbf{A}', \mathbf{v} > |.
\end{aligned}$$

We have assumed in Sect. 4.2.1 that $\tilde{\beta}_\nu \in \mathcal{C}^1_0(\overline{\mathbb{R}^+})$. Hence, $r \to \tilde{\beta}_\nu(r)r$ is Lipschitz continuous. Moreover, the Lipschitz constant $\tilde{L}$ satisfies

$$\tilde{L} = \sup_{r\in\mathrm{supp}(\tilde{\beta}_\nu)} |\tilde{\beta}'_\nu(r)r + \tilde{\beta}_\nu(r)| \leq \hat{C}\|\tilde{\beta}_\nu\|_{\mathcal{C}^1_0(\overline{\mathbb{R}^+_0})}. \tag{4.49}$$

We infer

$$
\begin{aligned}
| < \tilde{\mathcal{M}}(\mathbf{A}_s) - \tilde{\mathcal{M}}(\mathbf{A}), \mathbf{v} > | &\leq 3\tilde{L}|(\mathbf{curl}\,(\mathbf{A}_s - \mathbf{A}), \mathbf{curl}\,\mathbf{v})_{D_{\mathrm{C}}}| \\
&\leq \tilde{C}\|\mathbf{curl}\,(\mathbf{A}_s - \mathbf{A})\|_{L^\infty(D)^3}\|\mathbf{v}\|_{W_{\mathrm{st}}(D)},
\end{aligned}
$$

as the Lipschitz constant of $\mathbf{r} \mapsto \tilde{\beta}_\nu(|\mathbf{r}|)\mathbf{r}$ is $3\tilde{L}$. Using (4.38) we obtain

$$
\begin{aligned}
| < \mathcal{M}_{\mathrm{L}}\,\mathbf{A}' - \mathcal{M}_{\mathrm{L}}(\overline{\mathbf{A}_s\mathbf{A}})\mathbf{A}', \mathbf{v} > | &= \left| \int_0^1 a'(\mathbf{A}; \mathbf{A}', \mathbf{v}) - a'(\mathbf{A} + t(\mathbf{A}_s - \mathbf{A}); \mathbf{A}', \mathbf{v})\,\mathrm{d}t \right| \\
&\leq C_{\mathbf{h}}'\|\mathbf{curl}\,(\mathbf{A}_s - \mathbf{A})\|_{L^\infty(D)^3}\|\mathbf{A}'\|_{W_{\mathrm{st}}(D)}\|\mathbf{v}\|_{W_{\mathrm{st}}(D)} \\
&\leq C_{\mathbf{h}}'\tilde{C}\|\mathbf{curl}\,(\mathbf{A}_s - \mathbf{A})\|_{L^\infty(D)^3}\|\tilde{\beta}_\nu\|_{C(\overline{\mathbb{R}_0^+})}\|\mathbf{v}\|_{W_{\mathrm{st}}(D)},
\end{aligned}
$$

as $\|\mathbf{A}'\|_{W_{\mathrm{st}}(D)} \leq \tilde{C}\|\tilde{\beta}_\nu\|_{C(\overline{\mathbb{R}_0^+})}$ due to (4.42). Finally, we infer

$$
\|\mathbf{A}_s - \mathbf{A} - s\mathbf{A}'\|_{W_{\mathrm{st}}(D)} \leq \bar{C}s\|\tilde{\beta}_\nu\|_{C_0^1(\overline{\mathbb{R}_0^+})}\|\mathbf{curl}\,(\mathbf{A}_s - \mathbf{A})\|_{L^\infty(D)^3}. \tag{4.50}
$$

Dividing by $s\|\tilde{\beta}_\nu\|_{C_0^1(\overline{\mathbb{R}_0^+})}$ we see that $\mathbf{A}'$ can be identified with the Fréchet derivative as $\|\mathbf{curl}\,(\mathbf{A}_s - \mathbf{A})\|_{L^\infty(D)^3} \to 0$ for $s \to 0$. □

4.4.2 Interface Sensitivity

For the purpose of first order shape calculus we assume in this section that $\Gamma_{\mathrm{I},s}$ is of class $\mathcal{C}^{1,1}$. Here, the state equation reads, find $\mathbf{A}_s \in W_{\mathrm{st}}(D)$ such that

$$
(\nu_{\mathrm{C}}(|\mathbf{curl}\,\mathbf{A}_s|)\mathbf{curl}\,\mathbf{A}_s, \mathbf{curl}\,\mathbf{v})_{D_{\mathrm{C},s}} + (\nu_0\mathbf{curl}\,\mathbf{A}_s, \mathbf{curl}\,\mathbf{v})_{D_{\mathrm{E},s}} = (\mathbf{J}, \mathbf{v})_D \tag{4.51}
$$

for all $\mathbf{v} \in W_{\mathrm{st}}(D)$. Before investigating derivatives in this case, as the domain of definition of $\mathbf{A}$ is subject to variations, the precise meaning of $\mathbf{A}'$ should be clarified. We adapt the following definition of the shape derivative from [38].

Definition 4.1 (*Shape Derivative*) The shape derivative $\mathbf{A}'$ of $\mathbf{A}_s$ subject to (4.51) is formally defined as

$$
\mathbf{A}' := \lim_{s\to 0} \frac{\mathbf{A}_s(\mathbf{x}) - \mathbf{A}(\mathbf{x})}{s}, \tag{4.52}
$$

for all $\mathbf{x} \in (D_{\mathrm{E},s} \cap D_{\mathrm{E}}) \cup (D_{\mathrm{C},s} \cap D_{\mathrm{C}})$.

This section is devoted to characterize $\mathbf{A}'$ through a boundary value problem.

Remark 4.2 In shape optimization, $\mathbf{A}$ and Q, respectively, would be referred to as shape functions, as for all transformations $\mathcal{T}_s$, such that $\mathcal{T}_s(\Gamma_\mathrm{I}) = \Gamma_\mathrm{I}$,

$$\mathbf{A}[\mathcal{T}_s(\Gamma_\mathrm{I})] = \mathbf{A}[\Gamma_\mathrm{I}] \tag{4.53}$$

holds. Hence, the derivative (4.34) can be identified with the shape gradient and all tools developed in this context, see, e.g., [15] can be applied.

A main tool for shape sensitivity analysis is the structure theorem [15]. Here, we state a formulation given in the context of differential forms, applied to domain functionals, [42, Theorem 1]:

Theorem 4.1 (Structure Theorem) *For a domain $D \subset \mathbb{R}^3$ of class $\mathcal{C}^k$, $k \geq 1$ the domain functional defined as*

$$\hat{Q}(D) = \int_D u \, \mathrm{d}\mathbf{x} \tag{4.54}$$

where $u \in \mathcal{C}^k(\overline{D})$, is shape differentiable with shape gradient

$$\hat{Q}'(D) = \int_{\partial D} \mathcal{V}_n u \, \mathrm{d}\mathbf{x}, \tag{4.55}$$

where $\mathcal{V}_n = \boldsymbol{\mathcal{V}} \cdot \mathbf{n}$.

The shape derivative of $\mathbf{A}_s$ is characterized in the following lemma, where we recall that the outer unit normal $\mathbf{n}$, as well as the plus sign in the respective jump operators, refer to the domain D_C. Also, on a surface S the curl gives rise to a vector and scalar operator defined as

$$\mathbf{curl}_S \, u = \mathbf{grad} \, u \times \mathbf{n}, \tag{4.56}$$

$$\mathrm{curl}_S \, \mathbf{u} = \mathbf{curl} \, \mathbf{u} \cdot \mathbf{n}, \tag{4.57}$$

respectively, see also [43]. For simplicity, we also introduce the difference of a vector or scalar at a surface S as

$$[\mathbf{u}]_S := \mathbf{u}^+ - \mathbf{u}^-. \tag{4.58}$$

Lemma 4.1 *Let $\mathbf{A}_s$ be subject to (4.51) with $\boldsymbol{\mathcal{V}}$ as described in Sect. 4.2.3 with $k = 1$. The shape derivative $\mathbf{A}' \in \mathcal{H}_0(\mathbf{curl}, D_\mathrm{C}) \cap \mathcal{H}_0(\mathbf{curl}, D_\mathrm{E}) \cap \mathcal{H}(\mathrm{div}\, 0, D)$, satisfies*

$$(\nu_\mathrm{d}(\cdot, \mathbf{curl}\,\mathbf{A})\mathbf{curl}\,\mathbf{A}', \mathbf{curl}\,\mathbf{v})_D = -(\mathcal{V}_n[\nu]_{\Gamma_\mathrm{I}} \mathrm{curl}_{\Gamma_\mathrm{I}} \mathbf{A}, \mathrm{curl}_{\Gamma_\mathrm{I}} \mathbf{v})_{\Gamma_\mathrm{I}}, \tag{4.59}$$

for all $\mathbf{v} \in \mathcal{C}_0^\infty(D)^3$.

Proof As in the proof of Proposition 4.1 we define, for the ease of exposition,

$$< \mathcal{M}_s \mathbf{u}, \mathbf{v} >:= (\nu_\mathrm{C}(|\mathbf{curl}\,\mathbf{u}|)\mathbf{curl}\,\mathbf{u}, \mathbf{curl}\,\mathbf{v})_{D_{\mathrm{C},s}} + (\nu_0 \mathbf{curl}\,\mathbf{u}, \mathbf{curl}\,\mathbf{v})_{D_{\mathrm{E},s}} \tag{4.60}$$

for $\mathbf{u}, \mathbf{v} \in W_{\mathrm{st}}(D)$, where $\mathcal{M}_0 = \mathcal{M}$. We start from the perturbed and unperturbed state equation

$$< \mathcal{M}_s \mathbf{A}_s, \mathbf{v} > = (\mathbf{J}, \mathbf{v})_{D_\mathbf{J}}, \tag{4.61}$$

$$< \mathcal{M} \mathbf{A}, \mathbf{v} > = (\mathbf{J}, \mathbf{v})_{D_\mathbf{J}}. \tag{4.62}$$

We recall that $D_\mathbf{J} \cap D_{\mathrm{HA}} = \emptyset$ and hence the right-hand sides in the previous expressions remain unchanged. Subtracting both equations, as well as adding and subtracting $< \mathcal{M}_s \mathbf{A}, \mathbf{v} >$, we obtain

$$< \mathcal{M}_s \mathbf{A}_s - \mathcal{M}_s \mathbf{A}, \mathbf{v} > = < \mathcal{M} \mathbf{A} - \mathcal{M}_s \mathbf{A}, \mathbf{v} > . \tag{4.63}$$

For the right-hand side of the previous equation, dividing by s and taking the limit $s \to 0$, the structure theorem yields

$$\begin{aligned} \lim_{s \to 0} \frac{< \mathcal{M} \mathbf{A} - \mathcal{M}_s \mathbf{A}, \mathbf{v} >}{s} &= \left((\mathcal{V}_n \nu_0 \mathbf{curl}\, \mathbf{A}, \mathbf{curl}\, \mathbf{v})_{\Gamma_\mathrm{I}} - (\mathcal{V}_n \nu_\mathrm{C}(|\mathbf{curl}\, \mathbf{A}|) \mathbf{curl}\, \mathbf{A}, \mathbf{curl}\, \mathbf{v})_{\Gamma_\mathrm{I}} \right) \\ &= -(\mathcal{V}_n [\mathbf{h}(\cdot, \mathbf{curl}\, \mathbf{A})]_{\Gamma_\mathrm{I}}, \mathbf{curl}\, \mathbf{v})_{\Gamma_\mathrm{I}}. \end{aligned}$$

Equivalently for the left-hand side of (4.63), we obtain by Lemma 3.9

$$\lim_{s \to 0} \frac{\mathbf{h}(\cdot, \mathbf{curl}\, \mathbf{A}_s) - \mathbf{h}(\cdot, \mathbf{curl}\, \mathbf{A})}{s} = \boldsymbol{\nu}_\mathrm{d}(\cdot, \mathbf{curl}\, \mathbf{A}) \mathbf{curl}\, \mathbf{A}', \tag{4.64}$$

both on D_E and D_C and we conclude that $\mathbf{A}'$ is subject to

$$(\boldsymbol{\nu}_\mathrm{d}(\cdot, \mathbf{curl}\, \mathbf{A}) \mathbf{curl}\, \mathbf{A}', \mathbf{curl}\, \mathbf{v})_D = -(\mathcal{V}_n [\mathbf{h}(\cdot, \mathbf{curl}\, \mathbf{A})]_{\Gamma_\mathrm{I}}, \mathbf{curl}\, \mathbf{v})_{\Gamma_\mathrm{I}}. \tag{4.65}$$

Finally, on Γ_I, some direct manipulations yield

$$\mathbf{h}^+ - \mathbf{h}^- \underbrace{=}_{(4.33c)} \mathbf{n} \cdot (\mathbf{h}^+ - \mathbf{h}^-) \mathbf{n} \tag{4.66}$$

$$= (\nu^+ \mathbf{n} \cdot \mathbf{curl}\, \mathbf{A}^+ - \nu^- \mathbf{n} \cdot \mathbf{curl}\, \mathbf{A}^-) \mathbf{n} \tag{4.67}$$

$$= ([\nu]_{\Gamma_\mathrm{I}} \mathbf{n} \cdot \mathbf{curl}\, \mathbf{A}) \mathbf{n} \tag{4.68}$$

$$= ([\nu]_{\Gamma_\mathrm{I}} \mathrm{curl}_{\Gamma_\mathrm{I}} \mathbf{A}) \mathbf{n}, \tag{4.69}$$

which yields the desired result. □

Remark 4.3 At this stage it is unclear whether the integral on the right-hand-side of (4.59) is well defined, as under the minimal regularity assumption $\mathbf{A} \in \mathcal{H}(\mathbf{curl}, D)$, $\mathrm{curl}_{\Gamma_\mathrm{I}} \mathbf{A} \in \mathcal{H}^{-1/2}(\Gamma_\mathrm{I})$, solely, see [44]. Establishing such a result, is beyond the scope of this work.

Equivalently the strong form of the boundary value problem for the shape derivative can be derived.

Proposition 4.3 *The shape derivative given by (4.59) can be characterized by means of the boundary value problem*

$$\mathbf{curl}\,\left(\nu_{\mathrm{d}}(\cdot, \mathbf{curl}\,\mathbf{A})\mathbf{curl}\,\mathbf{A}'\right) = 0, \qquad \text{in } D_{\mathrm{C}} \cup D_{\mathrm{E}}, \tag{4.70a}$$

$$[\![\mathbf{A}']\!]_{\Gamma_{\mathrm{I}}} = \mathcal{V}_n \left(\mathrm{S}\mathbf{n} \times \mathbf{n}[A_n]_{\Gamma_{\mathrm{I}}} - [\![\partial_{\mathbf{n}}\mathbf{A}]\!]_{\Gamma_{\mathrm{I}}}\right), \quad \text{on } \Gamma_{\mathrm{I}}, \tag{4.70b}$$

$$[\![\nu_{\mathrm{d}}(\cdot, \mathbf{curl}\,\mathbf{A}')]\!]_{\Gamma_{\mathrm{I}}} = -\mathbf{curl}_{\,\Gamma_{\mathrm{I}}}\left(\mathcal{V}_n[\nu]_{\Gamma_{\mathrm{I}}}\mathrm{curl}_{\,\Gamma_{\mathrm{I}}}\mathbf{A}\right), \quad \text{on } \Gamma_{\mathrm{I}}, \tag{4.70c}$$

$$\mathbf{A}' \times \mathbf{n} = 0, \qquad \text{on } \partial D, \tag{4.70d}$$

$$\mathrm{div}\,\mathbf{A}' = 0, \qquad \text{in } D, \tag{4.70e}$$

where $\partial_n\mathbf{A} := (D\mathbf{A})\mathbf{n}$ *and* $\mathrm{S} = D\mathbf{n}$ *is the Weingarten map.*

Proof Equations (4.70d) and (4.70e) follow as the shape derivative is defined as the limit of a difference quotient. For the remaining identities, following [38], the starting point is

$$(\nu_{\mathrm{d}}(\cdot, \mathbf{curl}\,\mathbf{A})\mathbf{curl}\,\mathbf{A}', \mathbf{curl}\,\mathbf{v})_D = -(\mathcal{V}_n[\nu]_{\Gamma_{\mathrm{I}}}\mathrm{curl}_{\,\Gamma_{\mathrm{I}}}\mathbf{A}, \mathrm{curl}_{\,\Gamma_{\mathrm{I}}}\mathbf{v})_{\Gamma_{\mathrm{I}}}. \tag{4.71}$$

Testing with $\mathbf{v} \in \mathcal{C}_0^{\infty}(D_{\mathrm{C}})^3$ and $\mathbf{v} \in \mathcal{C}_0^{\infty}(D_{\mathrm{E}})^3$, respectively, we readily derive (4.70a) by performing integration by parts. Testing with $\mathbf{v} \in \mathcal{C}^{\infty}(\mathbb{R}^3)^3$ on Γ_{I} in turn and using the integration by parts formula

$$(\mathbf{curl}_{\,\Gamma_{\mathrm{I}}}\mathbf{u}, \mathbf{v})_{\Gamma_{\mathrm{I}}} = (\mathbf{u}, \mathrm{curl}_{\,\Gamma_{\mathrm{I}}}\mathbf{v})_{\Gamma_{\mathrm{I}}}, \tag{4.72}$$

we obtain (4.70c). Following [38] we start with a variational characterization of the jump of the trace. As $\mathbf{A}_s \in W_{\mathrm{st}}(D)$, we have

$$([\![\mathbf{A}_s]\!]_{\Gamma_{\mathrm{I},s}}, \mathbf{v})_{\Gamma_{\mathrm{I},s}} = 0, \tag{4.73}$$

for all $\mathbf{v} \in \mathcal{C}^{\infty}(D)^3$ such that $\mathbf{v} \cdot \mathbf{n} = 0$ on $\Gamma_{\mathrm{I},s}$. Note that, as Γ_{I} is fixed, we can define $\mathbf{n}$ globally as a normal field for all $\Gamma_{\mathrm{I},s}$, independent of s, cf. [42, p. 13]. Then differentiating in (4.73) and applying [42, Lemma 3] yields

$$([\![\mathbf{A}']\!]_{\Gamma_{\mathrm{I}}}, \mathbf{v})_{\Gamma_{\mathrm{I}}} = -\int_{\Gamma_{\mathrm{I}}} \mathcal{V}_n(\partial_{\mathbf{n}}([\![\mathbf{A}]\!]_{\Gamma_{\mathrm{I}}} \cdot \mathbf{v}) + \mathrm{H}([\![\mathbf{A}]\!]_{\Gamma_{\mathrm{I}}} \cdot \mathbf{v}))\,\mathrm{d}\mathbf{x}, \tag{4.74}$$

where H is the additive curvature. Using $[\![\mathbf{A}]\!]_{\Gamma_{\mathrm{I}}} = 0$ and the vector analysis identity

$$\mathbf{grad}\,(\mathbf{u} \cdot \mathbf{v}) = D\mathbf{v}\mathbf{u} + D\mathbf{u}\mathbf{v} + \mathbf{u} \times \mathbf{curl}\,\mathbf{v} + \mathbf{v} \times \mathbf{curl}\,\mathbf{u} \tag{4.75}$$

we obtain

$$
\begin{aligned}
([\![\mathbf{A}']\!]_{\Gamma_I}, \mathbf{v})_{\Gamma_I} &= -(\mathcal{V}_n \mathbf{grad}\,([\![\mathbf{A}]\!]_{\Gamma_I} \cdot \mathbf{v}), \mathbf{n})_{\Gamma_I}, && (4.76)\\
&= -(\mathcal{V}_n (D[\![\mathbf{A}]\!]_{\Gamma_I}\mathbf{v} - \mathbf{curl}\,[\![\mathbf{A}]\!]_{\Gamma_I} \times \mathbf{v}), \mathbf{n})_{\Gamma_I}, && (4.77)\\
&= -\underbrace{(\mathcal{V}_n D[\![\mathbf{A}]\!]_{\Gamma_I}\mathbf{v}, \mathbf{n})_{\Gamma_I}}_{=0,\ [\![\mathbf{A}]\!]_{\Gamma_I}\cdot\mathbf{n}=0} - (\mathcal{V}_n \mathbf{curl}\,[\![\mathbf{A}]\!]_{\Gamma_I} \times \mathbf{n}, \mathbf{v})_{\Gamma_I}. && (4.78)
\end{aligned}
$$

By means of

$$
\mathbf{curl}\,(\mathbf{u} \times \mathbf{n}) = \mathbf{u}\mathrm{div}\,\mathbf{n} - \mathbf{n}\mathrm{div}\,\mathbf{u} + D\mathbf{u}\mathbf{n} - D\mathbf{n}\mathbf{u} \qquad (4.79)
$$

we further obtain

$$
\begin{aligned}
\mathbf{curl}\,[\![\mathbf{A}]\!]_{\Gamma_I} \times \mathbf{n} =& (D\mathbf{A}^+\mathbf{n}^+ + D\mathbf{A}^-\mathbf{n}^-) \times \mathbf{n} + (\mathbf{A}^+\mathrm{div}\,\mathbf{n}^+ + \mathbf{A}^-\mathrm{div}\,\mathbf{n}^-) \times \mathbf{n} && (4.80)\\
&- (D\mathbf{n}^+\mathbf{A}^+ + D\mathbf{n}^-\mathbf{A}^-) \times \mathbf{n} && (4.81)\\
=& [\![\partial_{\mathbf{n}}\mathbf{A}]\!]_{\Gamma_I} + \underbrace{\mathrm{div}\,\mathbf{n}[\![\mathbf{A}]\!]_{\Gamma_I}}_{=0} - D\mathbf{n}(\mathbf{A}^+ - \mathbf{A}^-) \times \mathbf{n}. && (4.82)
\end{aligned}
$$

The result follows by observing that $\mathbf{A}^+ - \mathbf{A}^- = [A_n]_{\Gamma_I}\mathbf{n}$ on Γ_I. □

Remark 4.4 In the literature, continuous and discrete approaches to sensitivity analysis are distinguished. Our approach, as presented in this section is referred to as continuous, since the differential equation is differentiated first and then discretized. This avoids computing mesh sensitivities, however we do not obtain the exact gradient of the discretized system. It is questionable which approach should be consider to be more exact. In any case, both approaches are known to be asymptotically identical [45]. For a comparison in terms of cost vs. accuracy we also refer to [46].

4.4.3 Source Current Sensitivity

For a source current $\boldsymbol{\beta}_{\mathbf{J}} \in U^{\mathbf{J}}_{\mathrm{adm}}$ the Fréchet derivative of the vector potential is stated in the following result. The proof directly follows from the fact that the solution depends linearly on $\boldsymbol{\beta}_{\mathbf{J}}$.

Proposition 4.4 *Let* $\boldsymbol{\beta}_{\mathbf{J}} \in U^{\mathbf{J}}_{\mathrm{adm}}$ *be given by* $\boldsymbol{\beta}_{\mathbf{J}} = \boldsymbol{\beta}_{\mathbf{J},0} + \tilde{\boldsymbol{\beta}}_{\mathbf{J}}$ *with* $\tilde{\boldsymbol{\beta}}_{\mathbf{J}} \in \mathcal{H}(\mathrm{div}\,0, D)$. *The mapping* $\tilde{\boldsymbol{\beta}}_{\mathbf{J}} \mapsto \mathbf{A}[\boldsymbol{\beta}_{\mathbf{J}} + \tilde{\boldsymbol{\beta}}_{\mathbf{J}}]$ *is Fréchet differentiable with derivative* $\delta\mathbf{A} \in W_{\mathrm{st}}(D)$, *subject to*

$$\begin{aligned}
\mathbf{curl}\ (\boldsymbol{\nu}_{\mathrm{d}}(\cdot, \mathbf{curl}\ \mathbf{A})\mathbf{curl}\ \delta\mathbf{A}) &= 0, && \text{in } D_{\mathrm{C}}, && (4.83a)\\
\mathbf{curl}\ (\nu_0 \mathbf{curl}\, \delta\mathbf{A}) &= \tilde{\boldsymbol{\beta}}_{\mathbf{J}}, && \text{in } D_{\mathrm{E}}, && (4.83b)\\
[\![\delta\mathbf{A}]\!]_{\Gamma_{\mathrm{I}}} &= 0, && \text{on } \Gamma_{\mathrm{I}}, && (4.83c)\\
[\![\boldsymbol{\nu}_{\mathrm{d}}(\cdot, \mathbf{curl}\, \mathbf{A})\mathbf{curl}\, \delta\mathbf{A}]\!]_{\Gamma_{\mathrm{I}}} &= 0, && \text{on } \Gamma_{\mathrm{I}}, && (4.83d)\\
\delta\mathbf{A} \times \mathbf{n} &= 0, && \text{on } \partial D, && (4.83e)\\
\operatorname{div} \delta\mathbf{A} &= 0, && \text{in } D. && (4.83f)
\end{aligned}$$

4.5 Sensitivity Analysis, Adjoint Approach

The adjoint approach to sensitivity analysis is very popular and standard in optimization. Its main advantage lies in the fact, that despite the dimension of the input parameter space only one additional equation, the adjoint equation, has to be solved. The definition of adjoint operators is originally based on linearity and ambiguities arise in a nonlinear setting [47]. A popular choice, adapted here, is to employ the linearized operator in the adjoint equation. More precisely, let the adjoint variable $\mathbf{p} \in W_{\mathrm{st}}(D)$ be subject to

$$(\boldsymbol{\nu}_{\mathrm{d}}(\cdot, \mathbf{curl}\, \mathbf{A})\mathbf{curl}\, \mathbf{v}, \mathbf{curl}\, \mathbf{p})_D = \int_{D_{\mathrm{obs}}} q_1(\mathbf{v}) + q_2(\mathbf{curl}\, \mathbf{v})\ \mathrm{d}\mathbf{x}, \tag{4.84}$$

for all $\mathbf{v} \in W_{\mathrm{st}}(D)$. The adjoint representation of the gradient is stated in the following result.

Proposition 4.5 *Let* $\mathbf{A}'$ *be given by (4.37), (4.70) and (4.83) for the perturbations* $\tilde{\beta}_\nu$, $\mathcal{V}$ *and* $\tilde{\boldsymbol{\beta}}_{\mathbf{J}}$, *respectively. Let* $\mathbf{p}$ *be the solution of the adjoint problem (4.84). We have the following adjoint representations for the gradients*

$$\begin{aligned}
Q' &= -(\tilde{\beta}_\nu(|\mathbf{curl}\, \mathbf{A}|)\mathbf{curl}\, \mathbf{A}, \mathbf{curl}\, \mathbf{p})_{D_{\mathrm{C}}}, && (4.85a)\\
Q' &= -(\mathcal{V}_n[\nu]_{\Gamma_{\mathrm{I}}} \mathrm{curl}_{\Gamma_{\mathrm{I}}} \mathbf{A}, \mathrm{curl}_{\Gamma_{\mathrm{I}}} \mathbf{p})_{\Gamma_{\mathrm{I}}}, && (4.85b)\\
Q' &= (\tilde{\boldsymbol{\beta}}_{\mathbf{J}}, \mathbf{p})_{D_{\mathrm{E}}}. && (4.85c)
\end{aligned}$$

Proof We give the proof for $\tilde{\beta}_\nu$, the results for $\tilde{\boldsymbol{\beta}}_{\mathbf{J}}$ and $\mathcal{V}$ follow in the same way. For $\tilde{\beta}_\nu$, we recall that the derivative is subject to

$$(\boldsymbol{\nu}_{\mathrm{d}}(\cdot, \mathbf{curl}\, \mathbf{A})\mathbf{curl}\, \mathbf{A}', \mathbf{curl}\, \mathbf{v})_D = -(\tilde{\beta}_\nu(|\mathbf{curl}\, \mathbf{A}|)\mathbf{curl}\, \mathbf{A}, \mathbf{curl}\, \mathbf{v})_{D_{\mathrm{C}}}. \tag{4.86}$$

Choosing $\mathbf{v} = \mathbf{p}$ and $\mathbf{A}'$ as test functions in the previous equation and the adjoint equation (4.84), respectively, we infer, using the symmetry of $\boldsymbol{\nu}_{\mathrm{d}}$ that

$$Q' = \int_{D_{\mathrm{obs}}} q_1(\mathbf{A}') + q_2(\mathbf{curl}\, \mathbf{A}')\ \mathrm{d}\mathbf{x} = -(\tilde{\beta}_\nu(|\mathbf{curl}\, \mathbf{A}|)\mathbf{curl}\, \mathbf{A}, \mathbf{curl}\, \mathbf{p})_{D_{\mathrm{C}}}. \tag{4.87}$$

□

4.6 Sensitivity Analysis for the Time Transient Case

We proceed by carrying out a direct and adjoint sensitivity analysis for the time-transient case, however, restricted to the material coefficient as an input. This has been addressed already in [4] in the two-dimensional case and the results are reported here, under minor simplification and supplementation. Here, details about function spaces will be omitted. Let the QoI, modeling, e.g., the time averaged inductance, be given as

$$Q = \int_{I_T} \int_{D_{\mathrm{obs}}} q(u) \, \mathrm{d}\mathbf{x} \, \mathrm{d}t, \tag{4.88}$$

where q is still assumed to be linear. We recall that the state equation reads as

$$\begin{aligned} \sigma(\cdot)\frac{\partial u}{\partial t} - \mathrm{div}\,(\nu(\cdot, |\mathbf{grad}\, u|)\mathbf{grad}\, u) &= J_z, && \text{in } I_T \times D_{\mathrm{C}} \cup D_{\mathrm{E}}, && (4.89\mathrm{a}) \\ u &= 0, && \text{on } \{0\} \times D, && (4.89\mathrm{b}) \\ u &= 0, && \text{on } I_T \times \Gamma_{\mathrm{D}}. && (4.89\mathrm{c}) \end{aligned}$$

Let $\partial_t u := \partial u / \partial t$. Applying parametrization and differentiation as outlined in Sect. 4.4, u' can be characterized as the solution of

$$\begin{aligned} \sigma_{\mathrm{C}} \partial_t u' - \mathrm{div}\,\left(\boldsymbol{\nu}_{\mathrm{d}}(\cdot, \mathbf{grad}\, u)\mathbf{grad}\, u'\right) &= \mathrm{div}\,\left(\tilde{\beta}_\nu(|\mathbf{grad}\, u|)\mathbf{grad}\, u\right), && \text{in } I_T \times D_{\mathrm{C}}, \\ -\mathrm{div}\,\left(\boldsymbol{\nu}_{\mathrm{d}}(\cdot, \mathbf{grad}\, u)\mathbf{grad}\, u'\right) &= 0, && \text{in } I_T \times D_{\mathrm{E}}, \\ u' &= 0, && \text{on } \{0\} \times D, \\ u' &= 0, && \text{on } I_T \times \Gamma_{\mathrm{D}}. \end{aligned} \tag{4.90}$$

We consider the adjoint problem

$$\begin{aligned} -\sigma(\cdot)\partial_t p - \mathrm{div}\,(\boldsymbol{\nu}_{\mathrm{d}}(\cdot, \mathbf{grad}\, u)\mathbf{grad}\, p) &= q, && \text{in } I_T \times D_{\mathrm{C}} \cup D_{\mathrm{E}}, && (4.91\mathrm{a}) \\ p &= p_{\mathrm{end}}, && \text{on } \{T\} \times D, && (4.91\mathrm{b}) \\ p &= 0, && \text{on } I_T \times \Gamma_{\mathrm{D}}, && (4.91\mathrm{c}) \end{aligned}$$

where p_{end} is a suitably chosen terminal value, unspecified for the moment. With the aid of the substitution $t \to -\tilde{t}$ the terminal value formulation (4.91) can be transformed into an initial value formulation on the interval $[-T, 0]$ and standard numerical techniques can be used to solve it. Furthermore, the stability of (4.91) is implied by the stability of the linearized state problem.

Proposition 4.6 *Let p be the solution of the adjoint equation (4.91) with $p_{\mathrm{end}} = 0$ and let the input parameter be given by $\beta_\nu = \beta_{\nu,0} + \tilde{\beta}_\nu$ as described in Sect. 4.2.1. The gradient of (4.88) is given by*

$$Q' = -\int_{I_T}\int_{D_C} \tilde{\beta}_\nu(|\mathbf{grad}\ u|)\mathbf{grad}\ u \cdot \mathbf{grad}\ p \ \ \mathrm{d}\mathbf{x}\ \mathrm{d}t. \quad (4.92)$$

Proof Differentiating (4.88) with respect to u and using the adjoint equation (4.91) we obtain

$$Q' = \int_{I_T}\int_{D} q(u')\ \mathrm{d}\mathbf{x}\ \mathrm{d}t = \int_{I_T}\int_{D} (-\sigma(\cdot)\partial_t p - \mathrm{div}\ (\boldsymbol{\nu}_\mathrm{d}(\cdot, \mathbf{grad}\ u)\mathbf{grad}\ p))u'\ \mathrm{d}\mathbf{x}\ \mathrm{d}t.$$

Integration by parts in time and using the divergence theorem yields

$$Q' = \int_{I_T}\int_{D_C} \sigma_C p \partial_t u'\ \mathrm{d}\mathbf{x}\ \mathrm{d}t - \int_{D_C} \sigma_C p_\mathrm{end} u'(T)\ \mathrm{d}\mathbf{x} + \int_{I_T}\int_{D} \boldsymbol{\nu}_\mathrm{d}(\cdot, \mathbf{grad}\ u)\mathbf{grad}\ p \cdot \mathbf{grad}\ u'\ \mathrm{d}\mathbf{x}\ \mathrm{d}t. \quad (4.93)$$

Using, for all $\mathbf{x} \in D_C$,

$$\sigma_C \partial_t u' = \mathrm{div}\ \left(\boldsymbol{\nu}_\mathrm{d}(\cdot, \mathbf{grad}\ u)\mathbf{grad}\ u'\right) + \mathrm{div}\ \left(\tilde{\beta}_\nu(|\mathbf{grad}\ u|)\mathbf{grad}\ u\right) \quad (4.94)$$

in (4.93) we arrive, after using the divergence theorem, at

$$Q' = -\int_{D_C} \sigma_C p_\mathrm{end} u'(T)\ \mathrm{d}\mathbf{x} - \int_{I_T}\int_{D_C} \tilde{\beta}_\nu(|\mathbf{grad}\ u|)\mathbf{grad}\ p \cdot \mathbf{grad}\ u\ \mathrm{d}\mathbf{x}\ \mathrm{d}t. \quad (4.95)$$

This yields the result, as we can freely choose $p_\mathrm{end} = 0$ to eliminate $u'(T)$, which is unknown. □

4.7 Conclusion

In this chapter we parametrized the model equation with respect to different kind of input parameters on the differential equation level and outlined possible discrete input representations. A continuity result was established which is considered to be a prerequisite for any kind of uncertainty quantification. Finally, detailed results on direct and adjoint sensitivity analysis were obtained for the static case and the two-dimensional time-transient case. These results are at the core of uncertainty propagation techniques presented in the following.

References

1. Hlaváček, I., Chleboun, J., Babuška, I.: Uncertain Input Data Problems and the Worst Scenario Method. Elsevier (2004)
2. Ramarotafika, R., Benabou, A., Clénet, S.: Stochastic modeling of soft magnetic properties of electrical steels, application to stators of electrical machines. IEEE Trans. Magn. **48**, 2573–2584 (2012)
3. Adams, R.A., Fournier, J.J.F.: Sobolev Spaces, vol. 140. Academic Press (2003)
4. Römer, U., Schöps, S., Weiland, T.: Approximation of moments for the nonlinear magnetoquasistatic problem with material uncertainties. IEEE Trans. Magn. **50**(2) (2014)
5. Bartel, A., De Gersem, H., Hülsmann, T., Römer, U., Schöps, Sebastian, Weiland, Thomas: Quantification of uncertainty in the field quality of magnets originating from material measurements. IEEE Trans. Magn. **49**, 2367–2370 (2013)
6. Bartel, A., Hülsmann, T., Kühn, J., Pulch, R., Schöps, S.: Influence of measurement errors on transformer inrush currents using different material models. IEEE Trans. Magn. **50**(2), 485–488 (2014)
7. Brauer, J.R.: Simple equations for the magnetization and reluctivity curves of steel. IEEE Trans. Magn. **11**(1), 81–81 (1975)
8. Włodarski, Z.: Analytical description of magnetization curves. Phys. B: Condens. Matter **373**(2), 323–327 (2006)
9. Cimrák, I.: Material and shape derivative method for quasi-linear elliptic systems with applications in inverse electromagnetic interface problems. SIAM J. Numer. Anal. **50**(3), 1086–1110 (2012)
10. Fritsch, F.N., Carlson, R.E.: Monotone piecewise cubic interpolation. SIAM J. Numer. Anal. **17**(2), 238–246 (1980)
11. Heise, B.: Analysis of a fully discrete finite element method for a nonlinear magnetic field problem. SIAM J. Numer. Anal. **31**(3), 745–759 (1994)
12. Reitzinger, S., Kaltenbacher, B., Kaltenbacher, M.: A note on the approximation of B-H curves for nonlinear computations. Technical Report 02-30, SFB F013, Johannes Kepler University Linz, Austria (2002)
13. Römer, U., Schöps, S., Weiland, T.: Stochastic modeling and regularity of the nonlinear elliptic curl-curl equation. SIAM/ASA J Uncertainty Quantification (in press)
14. Pechstein, C., Jüttler, B.: Monotonicity-preserving interproximation of B-H curves. J. Comput. Appl. Math. **196**(1), 45–57 (2006)
15. Delfour, M.C., Zolésio, J.-P.: Shapes and Geometries: Metrics, Analysis, Differential Calculus, and Optimization, 1st edn. SIAM (2001)
16. Sokolowski, J., Zolésio, J.-P.: Introduction to Shape Optimization. Springer (1992)
17. Harbrecht, H., Schneider, R., Schwab, C.: Sparse second moment analysis for elliptic problems in stochastic domains. Numer. Math. **109**(3), 385–414 (2008)
18. Murat, F., Simon, J.: Etude de problèmes d'optimal design. In: Optimization Techniques Modeling and Optimization in the Service of Man Part 2. Springer (1976), pp. 54–62
19. Delfour, M.C., Zolésio, J.-P.: Structure of shape derivatives for nonsmooth domains. J. Funct. Anal. **104**(1), 1–33 (1992)
20. Eppler, K.: Optimal shape design for elliptic equations via bie-methods. Appl. Math. Comput. Sci. **10**(3), 487–516 (2000)
21. Harbrecht, H.: On output functionals of boundary value problems on stochastic domains. Math. Methods Appl. Sci. **33**(1), 91–102 (2010)
22. Cohen, E., Martin, T., Kirby, R.M., Lyche, T., Riesenfeld, R.F.: Analysis-aware modeling: understanding quality considerations in modeling for isogeometric analysis. Comput. Methods Appl. Mech. Eng. **199**(5), 334–356 (2010)
23. Hughes, T.J.R., Cottrell, J.A.: Isogeometric analysis: cad, finite elements, nurbs, exact geometry and mesh refinement. Comput. Methods Appl. Mech. Eng. **194**(39), 4135–4195 (2005)
24. Manh, N.D., Evgrafov, A., Gersborg, A.R., Gravesen, J.: Isogeometric shape optimization of vibrating membranes. Comput. Methods Appl. Mech. Eng. **200**(13), 1343–1353 (2011)

25. Cho, S., Ha, S.-H.: Isogeometric shape design optimization: exact geometry and enhanced sensitivity. Struct. Multidiscipl. Optim. **38**(1), 53–70 (2009)
26. Nguyen, D.-M., Evgrafov, A., Gravesen, J., Lahaye, D.: Iso-geometric shape optimization of magnetic density separators. COMPEL: Int. J. Comput. Math. Electr. Electron. Eng. **33**(4), 24–24 (2014)
27. Schöps, S., De Gersem, H., Weiland, T.: Winding functions in transient magnetoquasistatic field-circuit coupled simulations. COMPEL: Int. J. Comput. Math. Electr. Electron. Eng. **32**(6), 2063–2083 (2013)
28. Bedrosian, G.: A new method for coupling finite element field solutions with external circuits and kinematics. IEEE Trans. Magn. **29**(2), 1664–1668 (1993)
29. Im, C.-H., Kim, H.-K., Jung, H.-K.: Novel technique for current density distribution analysis of solidly modeled coil. IEEE Trans. Magn. **38**(2), 505–508 (2002)
30. Bossavit, A.: Edge elements for magnetostatics. Int. J. Numer. Modell.-Electron. Netw. Devices Fields **9**(1), 19–34 (1996)
31. Schöps, S.: Multiscale modeling and multirate time-integration of field/circuit coupled problems. PhD thesis, Katholieke Universiteit Leuven (2011)
32. Durand, S., Cimrák, I., Sergeant, P.: Adjoint variable method for time-harmonic Maxwell equations. COMPEL: Int. J. Comput. Math. Electr. Electron. Eng. **28**(5), 1202–1215 (2009)
33. Lukáš, D.: On solution to an optimal shape design problem in 3-dimensional linear magnetostatics. Appl. Math. **49**(5), 441–464 (2004)
34. Park, Il-H, Coulomb, J.L., Hahn, S.: Design sensitivity analysis for nonlinear magnetostatic problems by continuum approach. J. Phys. III **2**(11), 2045–2053 (1992)
35. Kim, D.-H., Lee, S.-H., Park, Il-H, Lee, J.-H.: Derivation of a general sensitivity formula for shape optimization of 2-d magnetostatic systems by continuum approach. IEEE Trans. Magn. **38**(2), 1125–1128 (2002)
36. Kim, D.-H., Ship, K.S., Sykulski, J.K.: Applying continuum design sensitivity analysis combined with standard em software to shape optimization in magnetostatic problems. IEEE Trans. Magn. **40**(2), 1156–1159 (2004)
37. Park, Il-H, Kwak, I.-G., Lee, H.-B., Hahn, S., Lee, Ki-Sik: Design sensitivity analysis for transient eddy current problems using finite element discretization and adjoint variable method. IEEE Trans. Magn. **32**(3), 1242–1245 (1996)
38. Harbrecht, H., Li, J.: First order second moment analysis for stochastic interface problems based on low-rank approximation. ESAIM. Math. Modell. Numer. Anal. **47**(05), 1533–1552 (2013)
39. Harbrecht, H.: A finite element method for elliptic problems with stochastic input data. Appl. Numer. Math. **60**(3), 227–244 (2010)
40. Harbrecht, H., Peters, M., Siebenmorgen, M.: Combination technique based k-th moment analysis of elliptic problems with random diffusion. J. Comput. Phys. **252**, 128–141 (2013)
41. Becker, R., Rannacher, R.: An optimal control approach to a posteriori error estimation in finite element methods. Acta Numer. **2001**(10), 1–102 (2001)
42. Hiptmair, R., Li, J.: Shape derivatives in differential forms I: an intrinsic perspective. Ann. Mate. Pura Appl. **192**(6), 1077–1098 (2013)
43. Buffa, A., Ciarlet, P.: On traces for functional spaces related to Maxwell's equations part II: Hodge decompositions on the boundary of Lipschitz polyhedra and applications. Math. Methods Appl. Sci. **24**(1), 31–48 (2001)
44. Buffa, A., Costabel, M., Sheen, D.: On traces for $H(curl, \omega)$ in Lipschitz domains. J. Math. Anal. Appl. **276**(2), 845–867 (2002)
45. Borggaard, J., Verma, A.: On efficient solutions to the continuous sensitivity equation using automatic differentiation. SIAM J. Sci. Comput. **22**(1), 39–62 (2000)
46. Nadarajah, S., Jameson, A.: A comparison of the continuous and discrete adjoint approach to automatic aerodynamic optimization. AIAA Paper **667**, 2000 (2000)
47. Estep, D.: A short course on duality, adjoint operators, Green's functions, and a posteriori error analysis. Lect. Notes (2004)

Chapter 5
Uncertainty Quantification

This chapter is devoted to the modeling and propagation of uncertainties with emphasis on aleatory uncertainty. High-dimensional parametric models will be derived and techniques for their efficient approximation will be discussed and compared. To this end, the sensitivity analysis tools, derived in the previous chapter will be of central importance.

5.1 Uncertainty Modeling

In Sect. 4.2 we have parametrized the magnetostatic model for different kind of inputs. In each case, an infinite dimensional input parameter $\boldsymbol{\beta}$ together with finite dimensional approximations $\boldsymbol{\beta} \approx \boldsymbol{\beta}_M(\mathbf{y})$, $\mathbf{y} \in \Gamma \subset \mathbb{R}^M$, was identified. This should be considered as an intermediate step towards modeling input uncertainties due to the following reasons. No attention was paid so far on choosing $\mathbf{y}$ in such a way that M is small. However, this is a very important issue as each additional parameter will result in increased costs for uncertainty propagation. Moreover, possible correlations of the parameters have to be identified. Consider, e.g., the parametric shapes of Fig. 4.2. Perturbing the control points uniformly and independently, i.e., without any correlation, results in different shape variations for a different number of parameters. The same holds true for perturbations of f_{HB}. Uncorrelated variations in spline coefficients will probably lead to more oscillatory variations of f_{HB} as those obtained by means of closed-form models such as (4.8). To account for correlation, additional information in practice is needed concerning the variability under consideration, in the best case by means of a large number of measurements. The more information available, the more accurately the model output uncertainties can be predicted.

U. Römer, *Numerical Approximation of the Magnetoquasistatic Model with Uncertainties*, Springer Theses, DOI 10.1007/978-3-319-41294-8_5

5.1.1 Probabilistic Description of Uncertainties

A probabilistic setting is introduced by means of a probability space $(\Omega, \mathcal{F}, \mathrm{P})$, where Ω refers to the set of outcomes, $\mathcal{F}$ to the sigma algebra of events and P to a probability measure, respectively, see [1] for details. Upper case letters will be employed for real random variables, i.e., $Y : \Omega \to \mathbb{R}$. Random inputs now take the form $\boldsymbol{\beta} : \Omega \to U$, in particular the perturbed inputs $\tilde{\boldsymbol{\beta}}$ will be characterized as random fields. This means that, e.g., for the magnetic reluctivity, for each B, $\nu_{\mathrm{C}}(\omega, B)$ is a random variable. Or, equivalently $\nu_{\mathrm{C}} : \Omega \times \mathbb{R}_0^+ \to \mathbb{R}^+$. For more on random fields, see [2]. As in the parametric case, random inputs have to be admissible, expressed as follows.

Assumption 5.1 (*Stochastic Admissibility*) We call a random input stochastic admissible, if almost surely (a.s) $\boldsymbol{\beta}(\omega) \in U_{\mathrm{adm}}$, or, equivalently

$$\mathrm{P}(\omega \in \Omega \mid \boldsymbol{\beta}(\omega) \in U_{\mathrm{adm}}) = 1. \tag{5.1}$$

This constraint combined with the modeling of Sect. 4.2 has consequences of the type of admissible random fields as explained in the following Remark 5.1.

Remark 5.1 (Normal Random Fields) Many random variables or fields in practice are normally distributed. However, this gives rise to difficulties in the context of partial differential equations with random material coefficients [3, 4]. In our case, consider the function f_{HB} in a stochastic setting. The assumption $\nu(\omega) \in U_{\mathrm{adm}}^{\nu}$ a.s. implies, that with probability one their exist $C_1, C_2 > 0$ such that $C_1 \leq \partial_B f_{HB}(\omega, B) \leq C_2$. Suppose that f_{HB} was a normal random field, then its spatial derivative would equally be a normal random field and this condition would be violated. Strictly speaking, normally distributed $B - H$ curves would violate monotonicity or posses unbounded derivatives with probability greater than zero. A remedy would consist in employing truncated random variables or approximating normal random fields by means of Jacobi chaos as outlined in [5].

5.1.2 Karhunen-Loève Expansion

The content of this section is almost fully contained in our contribution [6, Sect. 3]. We propose a means of finding M uncorrelated random variables $\mathbf{Y}$ such that $\boldsymbol{\beta}(\omega) \approx \boldsymbol{\beta}_M(\mathbf{Y}(\omega))$, for $\omega \in \Omega$, i.e., we address the discretization of random fields. Let $g : \Omega \times I$, $I \subset \mathbb{R}$ be such that $g(\cdot, s)$ is square integrable for all $s \in I$. With this one-dimensional setting we address in particular a stochastic f_{HB} or one-dimensional shapes, see [7] for efficient schemes in the higher-dimensional case. Suppose further that the expected value and covariance function of g, i.e.,

$$\mathrm{E}[g](s) = \int_{\Omega} g(\omega, s)\mathrm{dP}, \tag{5.2}$$

$$\mathrm{Cov}[g](s,t) = \int_{\Omega} (g(\omega, s) - \mathrm{E}[g](s))(g(\omega, t) - \mathrm{E}[g](t))\mathrm{dP}, \tag{5.3}$$

are known, e.g., through measurements. For a continuous covariance kernel $\mathrm{Cov}[g]$, g can be represented by means of the Karhunen-Loève (KL) expansion

$$g(\omega, s) = \mathrm{E}[g](s) + \sum_{n=1}^{\infty} \sqrt{\lambda_n} b_n(s) Y_n(\omega), \tag{5.4}$$

see [8]. The Y_n are uncorrelated and centered random variables with unit variance and determined by

$$Y_n = \frac{1}{\sqrt{\lambda_n}} \int_{\Omega} (g(\omega, s) - \mathrm{E}[g](s)) b_n(s)\mathrm{d}s, \; n = 1, 2, \ldots \tag{5.5}$$

from measurements. For simplicity, we additionally assume that they are independent and bounded here. The $(\lambda_n, b_n)_{n=1}^{\infty}$ refer to the eigenvalues and eigenfunctions of the eigenvalue problem

$$\int_I \int_I \mathrm{Cov}[g](s,t) b_n(s) v(t)\mathrm{d}s\mathrm{d}t = \lambda_n \int_I b_n(s) v(s)\mathrm{d}s, \tag{5.6}$$

for all $v \in L^2(I)$ and $\mathrm{Cov}[g] \in L^2(I \times I)$. Studying properties of the operator associated to the covariance one can deduce that $\lambda_1 \geq \lambda_2 \geq \cdots \geq 0$ and that $(b_n)_{n=1}^{\infty}$ is an orthonormal basis of $L^2(I)$. We accomplish the task of random discretization by truncating the series (5.4) after M terms. The KL expansion is particularly appealing as it is optimal among all M-term approximations in the L^2-norm [7]. Additionally, the error is given by

$$\|g - g_M\|^2_{L^2(I \times \Omega)} = \sum_{n=M+1}^{\infty} \lambda_n, \tag{5.7}$$

which can be estimated in practice. Another popular choice is given by means of generalized Polynomial Chaos (gPC) [5]. The truncated KL expansion

$$g_M(\omega, s) = \mathrm{E}[g](s) + \sum_{n=1}^{M} \sqrt{\lambda_n} b_n(s) Y_n(\omega), \tag{5.8}$$

cannot directly be applied in numerical approximations as the eigenvalues and eigenfunctions, except for several simple covariance functions, are not known analytically. Hence, we resort to a finite element approximation based on B-splines, where we refer to Appendix B for a definition. More precisely, the Galerkin approximation of (5.6) reads, find $\lambda_{N,n} \neq 0$, and $b_{N,n} \in \mathcal{S}_N^{q,k}$, $n \geq 1$, such that

$$\int_I \int_I \mathrm{Cov}[g](s,t) b_{N,n}(s) v_N(t) \mathrm{d}s\mathrm{d}t = \lambda_{N,n} \int_I b_{N,n}(s) v_N(s) \mathrm{d}s, \tag{5.9}$$

for all $v_N \in \mathcal{S}_N^{q,k}$. In matrix notation, we obtain

$$\mathbf{K}\mathbf{b}_n = \lambda_n \mathbf{M}\mathbf{b}_n, \tag{5.10}$$

where $\mathbf{b}_n$ denotes the vector of degrees of freedom associated to $b_{N,n}$ and

$$K_{ij} = \int_I \int_I \mathrm{Cov}[g](s,t) b_{N,j}(s) b_{N,i}(t) \mathrm{d}s\mathrm{d}t, \quad M_{ij} = \int_I b_{N,i}(s) b_{N,j}(s) \mathrm{d}s. \tag{5.11}$$

With the numerical eigenpairs at hand and neglecting discretization effects on the Y_n, we can set

$$g_{M,N}(\omega, s) = \mathrm{E}[g](s) + \sum_{n=1}^{M} \sqrt{\lambda_{N,n}} b_{N,n}(s) Y_n(\omega). \tag{5.12}$$

For M fixed, the discretization error $g_M - g_{M,N}$ decays for $N \to \infty$ as the standard finite element error, whereas the truncation error $g - g_M$ is dictated by the decay of the eigenvalues and hence by the smoothness of the covariance kernel. Decay rates have been given in [9]. In particular for the popular choice of a Gaussian kernel

$$\mathrm{Cov}[g](s,t) = \mathrm{e}^{-(s-t)^2/L^2}, \tag{5.13}$$

where L refers to the correlation length (see Example 5.1), the eigenvalues decay faster than exponentially. In practice we can choose M, e.g., such that $\Psi_M > 0.95$, where

$$\Psi_M := \sum_{n=1}^{M} \lambda_n \Big/ \left(\sum_{n=1}^{\infty} \lambda_n \right). \tag{5.14}$$

We refer to [10] for a more detailed description of random input representations from measured data in terms of principal component analysis in combination with a posteriori error estimation.

Provided that an input β is stochastic admissible, cf. Assumption 5.1, one has to ensure that the same holds true for its Karhunen-Loève approximation β_M, e.g., monotonicity needs to be preserved for f_{HB}. To this end, the $L^2(\Omega \times I)$-convergence of β to β_M is not sufficient [7]. Consider the stochastic $B - H$ curve restricted to an interval, denoted $f_{HB} = f_{HB}|_I$, by abuse of notation. Omitting the spatial discretization error (assume N large), we consider the truncated Karhunen-Loève expansion $f_{HB,M}$ of f_{HB}. We require

$$\|f_{HB}^{(i)} - f_{HB,M}^{(i)}\|_{L^\infty(\Omega \times I)} \to 0 \tag{5.15}$$

for $i = 0, 1$, as $M \to \infty$, where we set $f^{(i)} := \partial_x^i f(\cdot, x)$. Under mild assumptions on the smoothness of the covariance [6], the eigenvalues decay fast enough, such that a.s. for $\lambda_n \neq 0$,

$$\|f_{HB}^{(i)}(\omega, \cdot) - f_{HB,M}^{(i)}(\omega, \cdot)\|_{L^\infty(I)} \leq C_\epsilon \sum_{n=M+1}^{\infty} \lambda_n^{1/2-\epsilon}, \tag{5.16}$$

holds with $\epsilon > 0$, [7, Theorem 2.24]. The constant C_ϵ additionally depends on the covariance and $|\Gamma|$. By choosing $M \geq M_0$ large enough, for all $s \in I$, we obtain

$$C_1 \leq f_{HB,M}^{(1)}(\omega, s) \leq C_2, \tag{5.17}$$

where $C_1, C_2 > 0$ and the same holds true for $\nu_{C,M}(\omega, s) := f_{HB,M}(\omega, s)/s$. By means of a suitable prolongation to $\mathbb{R}^+$ a discrete and admissible input has been obtained. Complementary to this theoretical result, a concrete example is considered in the following.

Example 5.1 The content of this example can be found completely in our work [6]. For simplicity, the subscript *HB* and all units are omitted. Adapting the setting of Sect. 4.2.2, a table of measurements

$$\{(\hat{s}_i, \hat{f}_{ij}),\ i = 1, \ldots, N^{\text{ms}},\ j = 1, \ldots, Q\} \tag{5.18}$$

for Q different material samples is given and assumed to be monotonic, i.e., $\hat{f}_{i_1 j} \leq \hat{f}_{i_2 j}$, for $i_1 \leq i_2$ and $j = 1, \ldots, Q$. The correlation function

$$k_f(s, t) = \frac{\text{Cov}_f(s, t)}{\sqrt{\text{Cov}_f(s, s)}\sqrt{\text{Cov}_f(t, t)}}, \tag{5.19}$$

is chosen to be approximated by the Gaussian kernel (5.13). Following [11], we introduce a scaling factor $\delta > 0$ and approximate the material relation as

$$f_M(\omega, s) = \text{E}_f(s) + \delta \sum_{n=1}^{M} \sqrt{\lambda_n} b_n(s) Y_n(\omega), \tag{5.20}$$

where the eigenvalues and eigenfunctions λ_n, b_n (subscript N omitted) are obtained by solving (5.9) with $\mathcal{S}_{30}^{3,1}$. For simplicity, the Y_n are modeled to be distributed uniformly as $\mathcal{U}(-\sqrt{3}, \sqrt{3})$. We choose $M = 3$ which assures $\Psi_M > 0.95$. Admissibility, i.e., monotonicity, is ensured by appropriately choosing δ as follows. We recall from [12] that a sufficient condition for a B-spline to be monotonic is that its coefficients are increasing. Here, a monotonic $\text{E}_f \in \mathcal{S}_{30}^{3,1}$ with coefficient vector $(\text{E}_{f,i})_{i=1}^{30}$ is simply obtained by projection from the interpolated sample mean $(\text{E}_{\hat{f}_i})$. Let $\eta_M(s) = \sum_{n=1}^{M} \sqrt{\lambda_n} b_n(s)$ and $\eta_{M,i}$ be obtained by replacing b_n with its i-th spline coefficients. Then, monotonicity can be assured by setting

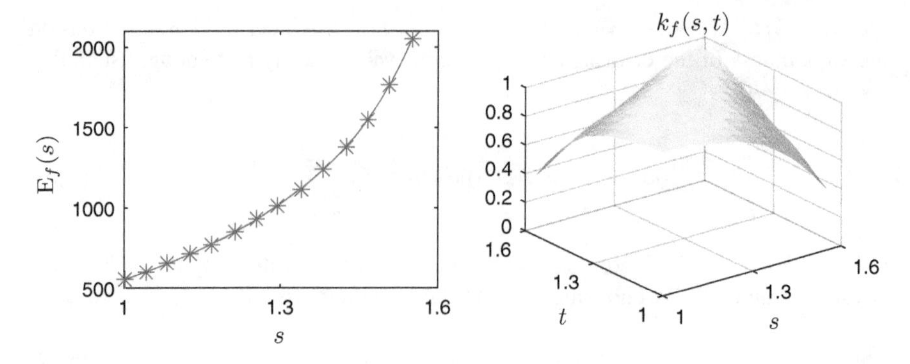

Fig. 5.1 Figure after [6]. Interpolated sample expected value E_f and correlation function k_f for data given in [13]

$$\delta < \min_{i=2,\ldots,N} \frac{E_{f,i} - E_{f,i-1}}{\sqrt{3}|\eta_{M,i} - \eta_{M,i-1}|}, \tag{5.21}$$

where we minimize only over those i with nonzero denominator.

In [13], measured data for an electrical machine,[1] representing the material properties from twenty-eight stator samples ($Q = 28$) from production chain was presented. Measurements are given for the interval $I = [1, 1.55]$ at $N^{\text{ms}} = 14$ equidistant points. In Fig. 5.1 both the expected value and the correlation function are depicted. In Fig. 5.2 sample realizations for different correlation lengths $L = 1/10$ and $L = 1/2$ are shown. In each case, the coefficient δ is chosen smaller than (5.21). Roughly speaking, two points on the interval I are correlated if their distance is smaller than the correlation length. Hence, a smaller correlation length corresponds to higher oscillatory sample trajectories. Increasing oscillations in the perturbation of the $B - H$ curve in turn demand for smaller perturbation amplitudes to ensure monotonicity. This is well illustrated in Fig. 5.2.

5.1.3 Stochastic Formulation and KL Modeling Error

In a next step we are going to describe a stochastic formulation of our model and establish its solvability. We refer to [11, 14] for stochastic formulations of elliptic differential equations, to [15] for a stochastic parabolic model and to [16, 17] for random domain, resp. interface, formulations. The stochastic magnetoquasistatic model reads, find the vector potential $\mathbf{A} : \Omega \times I_T \times D \to \mathbb{R}^3$ such that a.s.

[1]The simulation is based on the original data kindly provided by Stéphane Clénet.

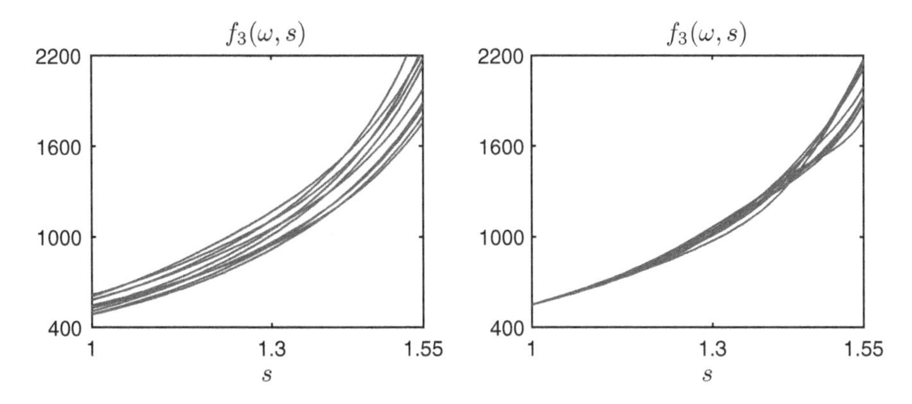

Fig. 5.2 Sample discretizations for the Karhunen-Loève approximation of the $B - H$ curve for different correlation lengths, see [6]. The correlation lengths and perturbation amplitudes are $L = 1/2$, $\delta = 2$ for the left and $L = 1/10$, $\delta = 1$ for the right, respectively, demonstrating the proportionality of increasing δ and L

$$\sigma(\omega, \mathbf{x})\frac{\partial \mathbf{A}}{\partial t}(\omega, t, \mathbf{x})+ \mathbf{curl}\ (\nu_\mathrm{C}(\omega, |\mathbf{curl}\,\mathbf{A}(\omega, t, \mathbf{x})|)\mathbf{curl}\,\mathbf{A}(\omega, t, \mathbf{x})) = 0, \text{ in } I_T \times D_\mathrm{C}(\omega), \tag{5.22a}$$

$$\mathbf{curl}\ (\nu_0\mathbf{curl}\,\mathbf{A}(\omega, t, \mathbf{x})) = \mathbf{J}(\omega, t, \mathbf{x}), \text{ in } I_T \times D_\mathrm{E}(\omega), \tag{5.22b}$$

$$[\![\mathbf{A}(\omega, t, \mathbf{x})]\!]_{\Gamma_\mathrm{I}(\omega)} = 0, \text{ on } I_T \times \Gamma_\mathrm{I}(\omega), \tag{5.22c}$$

$$[\![\nu(\omega, \mathbf{x}, |\mathbf{curl}\,\mathbf{A}(\omega, t, \mathbf{x})|)\mathbf{curl}\,\mathbf{A}(\omega, t, \mathbf{x}))]\!]_{\Gamma_\mathrm{I}(\omega)} = 0, \text{ on } I_T \times \Gamma_\mathrm{I}(\omega), \tag{5.22d}$$

$$\mathbf{A}(\omega, t, \mathbf{x}) \times \mathbf{n} = 0, \text{ on } I_T \times \Gamma_\mathrm{D}, \tag{5.22e}$$

$$\mathbf{A}(\omega, 0, \mathbf{x}) = \mathbf{A}_\mathrm{init}(\mathbf{x}), \text{ on } \{0\} \times D, \tag{5.22f}$$

$$\mathrm{div}\ \mathbf{A}(\omega, t, \mathbf{x}) = 0, \text{ in } I_T \times D_\mathrm{E}(\omega), \tag{5.22g}$$

$$\int_{\Gamma_{\mathrm{E},i}(\omega)} \mathbf{A}(t) \cdot \mathbf{n}\, \mathrm{d}\mathbf{x} = 0, \text{ on } I_T, \tag{5.22h}$$

holds. We write $\mathbf{A}(t) = \mathbf{A}(\cdot, t, \cdot)$ or $\mathbf{A}(\omega) = \mathbf{A}(\omega, \cdot, \cdot)$ for short, when no confusion is possible. The corresponding weak formulation is obtained by taking the expected value of (3.10), i.e., almost everywhere in I_T, find $\mathbf{A} \in L^2(\Omega) \otimes L^2(I_T, W(D))$, $\dot{\mathbf{A}}|_{D_\mathrm{C}} \in L^2(\Omega) \otimes L^2(I_T, L^2(D_\mathrm{C})^3)$, such that

$$\int_{D_\mathrm{C}} \mathrm{E}[\sigma_\mathrm{C}\dot{\mathbf{A}}(t) \cdot \mathbf{v}]\, \mathrm{d}\mathbf{x}+ \int_D \mathrm{E}[\nu(\cdot, \cdot, |\mathbf{curl}\,\mathbf{A}(t)|)\mathbf{curl}\,\mathbf{A}(t) \cdot \mathbf{curl}\,\mathbf{v}]\, \mathrm{d}\mathbf{x} = \int_D \mathrm{E}[\mathbf{J}(t) \cdot \mathbf{v}]\, \mathrm{d}\mathbf{x}, \tag{5.23}$$

for all $\mathbf{v} \in L^2(\Omega) \otimes W(D)$. Note that the stochastic interface generates randomness in the entire magnetic reluctivity as

$$\nu(\omega, \mathbf{x}, \cdot) = \nu_0 \chi_{D_{\mathrm{E}}(\omega)}(\mathbf{x}) + \nu_{\mathrm{C}}(\omega, \cdot) \chi_{D_{\mathrm{C}}(\omega)}(\mathbf{x}), \tag{5.24}$$

see [17] and also Remark 4.1. For inputs that are stochastic admissible, (5.22), resp. (5.23), is uniquely solvable a.s. and by the energy estimates as given, e.g., in [18] all moments of the solution of (5.23) exist, i.e., $\mathbf{A} \in L^p(\Omega) \otimes \mathcal{C}([0, T], W(D))$ for all $p > 0$. We proceed by deriving a stability result for the vector potential with respect to the input randomness. For simplicity, the result is given for the stochastic magnetostatic formulation and a random magnetic reluctivity, solely. The corresponding formulation reads, find $\mathbf{A} \in L^2(\Omega) \otimes W_{\mathrm{st}}(D)$ subject to

$$\mathbf{curl}\ (\nu_{\mathrm{C}}(\omega, |\mathbf{curl\,A}(\omega, \mathbf{x})|)\mathbf{curl\,A}(\omega, \mathbf{x})) = 0, \text{ in } D_{\mathrm{C}}(\omega), \tag{5.25a}$$
$$\mathbf{curl}\ (\nu_0 \mathbf{curl\,A}(\omega, \mathbf{x})) = \mathbf{J}(\mathbf{x}), \text{ in } D_{\mathrm{E}}(\omega), \tag{5.25b}$$
$$[\![\mathbf{A}(\omega, \mathbf{x})]\!]_{\Gamma_{\mathrm{I}}(\omega)} = 0, \text{ on } \Gamma_{\mathrm{I}}(\omega), \tag{5.25c}$$
$$[\![\nu(\omega, \mathbf{x}, |\mathbf{curl\,A}(\omega, \mathbf{x})|)\mathbf{curl\,A}(\omega, \mathbf{x}))]\!]_{\Gamma_{\mathrm{I}}(\omega)} = 0, \text{ on } \Gamma_{\mathrm{I}}(\omega), \tag{5.25d}$$
$$\mathbf{A}(\omega, \mathbf{x}) \times \mathbf{n} = 0, \text{ on } \Gamma_{\mathrm{D}}, \tag{5.25e}$$
$$\operatorname{div} \mathbf{A}(\omega, \mathbf{x}) = 0, \text{ in } D_{\mathrm{E}}(\omega), \tag{5.25f}$$

for all $\mathbf{v} \in L^2(\Omega) \otimes W_{\mathrm{st}}(D)$, cf. [15]. The error in the vector potential, when ν_{C} is approximated by means of $\hat{\nu}_{\mathrm{C}}$ in (5.25) is quantified in the following Proposition, adapted from our work [6, Proposition 1]. To lighten notation we write $\|\cdot\|_{W_{\mathrm{st}}(D)} = \|\cdot\|_{\mathcal{H}(\mathbf{curl}, D)}$ in the remaining part of this section.

Proposition 5.1 *Let* $\mathbf{A}$ *and* $\hat{\mathbf{A}}$ *be the weak solution of (5.25) for* ν_{C} *and* $\hat{\nu}_{\mathrm{C}}$*, respectively. Let Assumption 5.1 be satisfied for* ν_{C} *and* $\hat{\nu}_{\mathrm{C}}$*, in particular* $\hat{\nu}_{\mathrm{C}} \in [\hat{\nu}_{\min}, \nu_0]$*. Then we have a.s.*

$$\|\mathbf{A} - \hat{\mathbf{A}}\|_{W_{\mathrm{st}}(D)} \leq \|\nu_{\mathrm{C}} - \hat{\nu}_{\mathrm{C}}\|_{L^\infty(\mathbb{R}^+)} \frac{C_{\mathrm{F}} \|\mathbf{J}\|_2}{\nu_{\min} \hat{\nu}_{\min}}. \tag{5.26}$$

Proof As ν is subject to Assumption 5.1, the operator

$$< \mathcal{M}(\omega)\mathbf{u}, \mathbf{v} > := \int_D \nu(\omega, \cdot, |\mathbf{curl\,u}|)\mathbf{curl\,u}, \mathbf{curl\,v}, \tag{5.27}$$

where $\mathbf{u}, \mathbf{v}$ in $W_{\mathrm{st}}(D)$, is a.s. strongly monotone and we obtain

$$\nu_{\min} \|\mathbf{A} - \hat{\mathbf{A}}\|^2_{W_{\mathrm{st}}(D)} \leq < \mathcal{M}(\cdot)\mathbf{A} - \mathcal{M}(\cdot)\hat{\mathbf{A}}, \mathbf{A} - \hat{\mathbf{A}} > . \tag{5.28}$$

Let $\hat{\mathcal{M}}$ denote the operator obtained by replacing ν by means of $\hat{\nu}$ in (5.27), then a.s. $\mathcal{M}(\omega)\mathbf{A}(\omega) = \hat{\mathcal{M}}(\omega)\hat{\mathbf{A}}(\omega)$ and by the Cauchy-Schwarz inequality (5.28) can be recast as

$$\nu_{\min}\|\mathbf{A}-\hat{\mathbf{A}}\|_{W_{\mathrm{st}}(D)} \leq \left(\int_{D_\mathrm{C}} |(\hat{\nu}_\mathrm{C}(\cdot,|\mathbf{curl}\,\hat{\mathbf{A}}|)-\nu_\mathrm{C}(\cdot,|\mathbf{curl}\,\hat{\mathbf{A}}|))\mathbf{curl}\,\hat{\mathbf{A}}|^2\,\mathrm{d}\mathbf{x}\right)^{1/2} \tag{5.29}$$

$$\leq \|\hat{\nu}_\mathrm{C}-\nu_\mathrm{C}\|_{L^\infty(R^+)}\|\hat{\mathbf{A}}\|_{W_{\mathrm{st}}(D)}. \tag{5.30}$$

The result follows from $\|\hat{\mathbf{A}}\|_{W_{\mathrm{st}}(D)} \leq C_\mathrm{F}\|\mathbf{J}\|_2/\hat{\nu}_{\min}$. □

The previous result can be used, e.g., to quantify the error due to an approximation by a truncated Karhunen-Loève expansion. Assuming now, that M independent random variables $\mathbf{Y}$ have been identified, accurately describing the input randomness, it follows from the Doob-Dynkin Lemma that the solution can be written as $\mathbf{A}(\omega,t,\mathbf{x})=\mathbf{A}(\mathbf{Y}(\omega),t,\mathbf{x})$ [11], by abuse of notation. Then the stochastic formulation possesses a high-dimensional, deterministic equivalent. To this end, let ϱ be the joint probability density function

$$\varrho:\Gamma\to\mathbb{R}, \tag{5.31}$$

such that $\varrho(\mathbf{Y})=\varrho_1(Y_1)\varrho_2(Y_2)\ldots\varrho_M(Y_M)$. Then the expected value can be recast as

$$\mathrm{E}[\mathbf{Y}]=\int_\gamma \mathbf{y}\varrho(\mathbf{y})\mathrm{d}\mathbf{y}. \tag{5.32}$$

An equivalent formulation of (5.23) is given by, find $\mathbf{A}\in L^2_\varrho(\Gamma)\otimes L^2(I_T,W(D))$, $\dot{\mathbf{A}}_\mathrm{C}\in L^2_\varrho(\Gamma)\otimes L^2(I_T,L^2(D_\mathrm{C})^3)$, such that almost everywhere in I_T,

$$\int_{\Gamma\times D_\mathrm{C}} \sigma_\mathrm{C}\dot{\mathbf{A}}(t)\cdot\mathbf{v}\varrho\mathrm{d}\mathbf{y}\,\mathrm{d}\mathbf{x}+ \int_{\Gamma\times D} \nu(\cdot,\cdot,|\mathbf{curl}\,\mathbf{A}(t)|)\mathbf{curl}\,\mathbf{A}(t)\cdot\mathbf{curl}\,\mathbf{v}\varrho\mathrm{d}\mathbf{y}\,\mathrm{d}\mathbf{x}=\int_{\Gamma\times D}\mathbf{J}(t)\cdot\mathbf{v}\varrho\mathrm{d}\mathbf{y}\,\mathrm{d}\mathbf{x}, \tag{5.33}$$

holds for all $\mathbf{v}\in L^2_\varrho(\Gamma)\otimes W(D)$. The high-dimensional parametric equation (5.33) will be the starting point for several discretization schemes.

5.2 Model Dimension Reduction and Uncertainty Propagation

This section is concerned with the efficient approximation of the stochastic equation (5.23) or the high-dimensional deterministic equation (5.33) derived in the previous section. Depending on the type of formulation used, methods are frequently classified into stochastic and deterministic, respectively [5]. From this perspective, the Monte Carlo method, briefly recalled in Sect. 5.2.2, is a stochastic method, whereas the moment based perturbation method of Sect. 5.2.3 and the methods based on generalized polynomial chaos in Sect. 5.2.4 are referred to as deterministic. Moreover, one can often distinguish between schemes characterizing the systems variability around the mean value, where the primary aim is the computation of some statistical moments and schemes for the computation of failure probabilities or the worst-case analysis presented in Sect. 5.2.5.

Most of the methods presented here, with exception of the Monte-Carlo method and adjoint based methods, suffer from the curse of dimensionality, i.e., the accuracy vs. cost ratio becomes impracticable for large M. Indeed, approximating the function $\mathbf{A} : \Gamma \times I_T \times D \rightarrow \mathbb{R}^3$ within the deterministic formulation (5.33) requires discretization in $M + 3 + 1$ dimensions. Note that when different types of inputs are considered, M might become large very quickly, even if the Karhunen-Loève expansion is used for each input separately. To this end in a first step a technique for reducing the stochastic dimension is presented.

Several methods and schemes presented are well-established and mainly adapted here to the nonlinear model under consideration. More efforts are needed for the perturbation method of Sect. 5.2.3 as the results depend on the model under consideration. In this case most of the results have been given for linear models. Here, instead we consider the nonlinear material law as an input recalling our contribution [19]. As a further new contribution, a regularity analysis is carried out in Sect. 5.2.4, also with respect to the reluctivity as an input [6]. This is an important aspect, as it determines the convergence rates and hence the efficiency of the collocation-based polynomial chaos method.

5.2.1 Dimension Reduction

In this section a technique for dimension reduction is presented from [20] and the principles of global sensitivity analysis are sketched. Assume we are interested in integrating a multivariate function (e.g., the QoI) $\hat{Q} : \Gamma \rightarrow \mathbb{R}$. In particular the aim is to compute some of its moments, which might be costly, in particular for large M. In this case, using a High-Dimensional Model Representation (HDMR) expansion, a low-dimensional approximate model can be obtained. To this end, consider the cut-HDMR expansion

$$\hat{Q}(\mathbf{Y}) = \hat{Q}_0 + \sum_{i=1}^{M} \hat{Q}_i(Y_i) + \sum_{1 \leq i_1 < i_2 \leq M} \hat{Q}_{i_1,i_2}(Y_{i_1}, Y_{i_2}) + \cdots$$
$$+ \sum_{1 \leq i_1 < \cdots < i_s \leq M} \hat{Q}_{i_1 \ldots i_s}(Y_{i_1}, \ldots, Y_{i_s}) + \cdots + \hat{Q}_{12 \ldots M}(Y_1, \ldots, Y_M), \quad (5.34)$$

where the constant, uni- and bi-variate terms are given by

$$\hat{Q}_0 = \hat{Q}(\hat{\mathbf{Y}}), \tag{5.35}$$

$$\hat{Q}_i(Y_i) = \hat{Q}(\hat{Y}_i) - \hat{Q}_0, \quad i = 1, \ldots, M, \tag{5.36}$$

$$\hat{Q}_{ij}(Y_i, Y_j) = \hat{Q}(\hat{Y}_{ij}) - \hat{Q}_i(Y_i) - \hat{Q}_j(Y_j) - \hat{Q}_0, \quad 1 \leq i < j \leq M, \tag{5.37}$$

respectively and $\hat{\mathbf{Y}} \in \Gamma$ is referred to as anchor point. The notation $\hat{Y}_i$ expresses that all vector components, except for i, are identified with those of the anchor point. For a complete definition of the terms appearing in (5.34) we refer again to [20]. The expansion (5.34) is finite, exact and the different terms account for combined interaction effects. Note that in the context of analysis of variance (ANOVA), an expansion similar to (5.34) is employed, where the variance of each term with respect to the overall variance is referred to as global sensitivity [21]. Now in the case of the cut-HDMR, based on the observation that in practice, effects of order greater than two are often negligible, a truncated expansion

$$\hat{Q}(\mathbf{Y}) = \hat{Q}_0 + \sum_{i=1}^{M} \hat{Q}_i(Y_i) + \sum_{1 \leq i_1 < i_2 \leq M} \hat{Q}_{i_1,i_2}(Y_{i_1}, Y_{i_2}) \tag{5.38}$$

can be employed. This expansion could be readily used in the context of uncertainty propagation, however, very likely, not all of the terms will have the same effect. To reduce the overall computation cost we adapt Algorithm 3.1 of [20] to identify the most important inputs. It consists in using finite differences at different anchor points, in order to estimate the magnitudes of $\hat{Q}_0$, as well as the $\hat{Q}_i$ and $\hat{Q}_{ij}$, relative to each other. However, here, we only consider one anchor point and adapt the setting to our needs.

Algorithm 5.1 (*cut-HDMR*)

(A.1) define treshholds ϵ_1 and ϵ_2 to classify the importance of uni- and bi-variate terms
(A.2) choose the center of Γ as an anchor point $\hat{\mathbf{Y}}$
(A.3) choose $\hat{h}_i$ small with respect to $|\Gamma_i|$ for $i = 1, \ldots, M$
(A.4) evaluate $\hat{Q}(\hat{\mathbf{Y}})$
(A.5) evaluate $\hat{Q}(\hat{\mathbf{Y}} + \hat{h}_i \mathrm{e}_i)$ for $i = 1, \ldots, M$
(A.6) identify Y_i as important and include $\hat{Q}_i(Y_i)$ in the cut-HDMR expansion if

$$|\hat{Q}(\hat{\mathbf{Y}} + \hat{h}_i \mathrm{e}_i) - \hat{Q}(\hat{\mathbf{Y}})| > \epsilon_1 \tag{5.39}$$

(A.7) evaluate $\hat{Q}(\hat{\mathbf{Y}} + 2\hat{h}_i \mathrm{e}_i + 2\hat{h}_j \mathrm{e}_j)$ for $i, j = 1, \ldots, M, i < j$
(A.8) identify the interaction of Y_i and Y_j as important and include $\hat{Q}_{ij}(Y_i, Y_j)$ in the cut-HDMR expansion if

$$|\hat{Q}(\hat{\mathbf{Y}} + 2\hat{h}_i \mathrm{e}_i + 2\hat{h}_j \mathrm{e}_j)| > \epsilon_2 \tag{5.40}$$

Here, e_i refers to the unit normal vector in direction Γ_i. The reduced cut-HDMR expansion can then be used for estimating, e.g., the moments. However, any other representation taking into account the important inputs and interactions could be used instead. The overall cost for dimension reduction is given by $M + 1 + (M(M-1))/2$ [20]. Together with the cost of uncertainty propagation in terms of the reduced model, this needs to be smaller than the cost of directly propagating uncertainties by means of $\hat{Q}$.

5.2.2 Monte Carlo Sampling

The Monte Carlo method, see [22] and the references therein, is a widely used tool for high-dimensional integration and uncertainty propagation. Consider the setting of the previous section, where we still aim at integrating a square-integrable function $\hat{Q} : \Gamma \to \mathbb{R}$, e.g., to compute its variance. In its simplest form, referred to as classical Monte Carlo method here, N^{MC} random samples $(\mathbf{Y}_i)_{i=1}^{N^{\mathrm{MC}}}$ are drawn from Γ, $\hat{Q}$ is evaluated for each sample and the mean value is estimated as

$$\mathrm{E}[\hat{Q}] \approx \mathrm{E}_{\mathrm{MC}}[\hat{Q}] = \frac{1}{N^{\mathrm{MC}}} \sum_{i=1}^{N^{\mathrm{MC}}} \hat{Q}(\mathbf{Y}_i). \tag{5.41}$$

This procedure is easy to implement as only repetitive simulations of the same code for different inputs are needed, i.e., the method is non-intrusive. Additionally, the root-mean-square error can be estimated as

$$\sqrt{\mathrm{E}[|\mathrm{E}[\hat{Q}] - \mathrm{E}_{\mathrm{MC}}[\hat{Q}]|^2]} = \frac{\mathrm{Std}[\hat{Q}]}{\sqrt{N^{\mathrm{MC}}}}, \tag{5.42}$$

see [22], where Std refers to the standard deviation, i.e., $\mathrm{Std}^2 = \mathrm{Var}$. Relation (5.42) yields an asymptotically exact error estimator by replacing Std in turn with its Monte Carlo counterpart

$$\mathrm{Std}_{\mathrm{MC}}[\hat{Q}] = \left(\frac{1}{N^{\mathrm{MC}} - 1} \sum_{i=1}^{N^{\mathrm{MC}}} \left(\hat{Q}[\mathbf{Y}_i] - \mathrm{E}_{\mathrm{MC}}[\hat{Q}] \right)^2 \right)^{1/2}, \tag{5.43}$$

see [22]. A well-known drawback is the slow, though dimension independent, convergence rate of $1/\sqrt{N^{\mathrm{MC}}}$ in (5.42), which results in a prohibitively high computational cost. Hence, in our case, where each evaluation of $\hat{Q}$ requires the solution of a partial differential equation, the Monte Carlo method will be mainly used as a benchmark. Note, that still considerable research is devoted to overcome the slow convergence rate as for very high-dimensional models options are limited. We mention quasi-Monte Carlo [22] and in particular multilevel Monte Carlo techniques [23, 24] in this context.

5.2.3 *Perturbation Methods for the Statistical Moments*

Assuming small input uncertainties, we propose a deterministic perturbation approach to approximate the statistical moments of a QoI by means of a first order Taylor expansion. The main interest is a reduction of the computational cost for uncertainty propagation with the drawback of a possibly reduced accuracy due to linearization. The framework is adapted from our work [19], to approximate the k-th statistical moment of QoIs. The asymptotic accuracy of the first order scheme is mathematically and numerically analyzed by a two-dimensional example of a transformer, Example 5.2.

To fully characterize a random variable/field, its probability density function, or cumulative distribution function, must be determined. However, here we often content ourselves with some of its moments. Of particular importance are the expected value $\mathrm{E}[g]$ and the variance

$$\mathrm{Var}[g] = \int_{\Omega} (g(\omega) - \mathrm{E}[g])^2 \mathrm{dP}, \tag{5.44}$$

of a random field g. In a broader context, we are interested in it's k-th moment

$$\tilde{\mathrm{M}}^k[g] := \int_{\Omega} g^k(\omega)\, \mathrm{dP}, \tag{5.45}$$

for $k \in \mathbb{N}$. We also frequently employ the central moments M^k around the mean, where for $k = 2$ we obtain the variance. In practice, often a few moments are sufficient to characterize the system's uncertain behavior. For example, if the skewness and kurtosis related to $\tilde{\mathrm{M}}^3$ and $\tilde{\mathrm{M}}^4$, respectively, are sufficiently small a random variable can be assumed to be normally distributed. In this case the probability density function is completely determined by mean and variance. Here, we aim at efficiently computing the moments of QoIs subject to the magnetoquasistatic model with an uncertain material coefficient. When the input uncertainties are sufficiently small the so-called *first order second moment analysis* [25–28] has been proposed to approximate mean and variance of the solution by means of perturbation techniques. We also refer to [29], where the linear Hodge Laplacian was considered, which coincides with the magnetostatic model in a special case. Most of these works address linear models, with the exception of the abstract nonlinear framework [27]. Also emphasis is put on the moments of the solution. Here, we focus on a nonlinear setting and propose a method to compute the moments of some QoI building on the work [30] and our contribution [19], which was limited to moments of degree two. The main ingredients are a stochastic Taylor expansion and a spline model, as well as tools for high-dimensional integration. In the latter case we propose to use error controlled Monte Carlo integration as an easy-to-implement and dimension-independent quadrature procedure. We consider the vector potential formulation in a two-dimensional setting, noting that the principles of the method presented remain valid in three dimensions.

Let the reluctivity be given in the form $\nu_{C,s}(\omega,\cdot) = \nu_C(\cdot) + s\tilde{\nu}_C(\omega,\cdot)$, with $s > 0$ small, and where, without loss of generality $\mathrm{E}[\tilde{\nu}_C] = 0$. We introduce the stochastic Taylor expansion

$$\hat{Q}_s(\omega) := \hat{Q}(\nu_{C,s}(\omega)) = \hat{Q}(\nu_C) + s\delta\hat{Q}(\tilde{\nu}_C(\omega)) + \mathcal{O}(s^2). \tag{5.46}$$

We recall that $\delta\hat{Q}$ refers to the Fréchet derivative. Hence, (5.46) is not always justified by the theory of Sect. 4, as the geometry and the data in this section might not always be covered by the assumptions of Proposition 4.3.

An important observation is that the gradient is mean free, as stated in the following result adapted from [30].

Lemma 5.1 *Let* $\tilde{\nu}_C$ *be centered, i.e.,* $\mathrm{E}[\tilde{\nu}_C] = 0$, *then* $\mathrm{E}[\delta\hat{Q}] = 0$.

Proof We first note that due to the adjoint representation of the gradient (Proposition 4.13) we have with the adjoint solution p

$$\mathrm{E}[\delta\hat{Q}] = -\int_\Omega\int_{I_T}\int_{D_C} \tilde{\nu}_C(\cdot, |\mathbf{grad}\ u|)\mathbf{grad}\ u \cdot \mathbf{grad}\ p \ \mathrm{d}\mathbf{x}\,\mathrm{d}t\,\mathrm{dP} = \tag{5.47}$$

$$-\int_{I_T}\int_{D_C} \underbrace{\left(\int_\Omega \tilde{\nu}_C(\cdot, |\mathbf{grad}\ u|)\,\mathrm{dP}\right)}_{=0,\ \mathrm{E}[\tilde{\nu}_C]=0} \mathbf{grad}\ u \cdot \mathbf{grad}\ p \ \mathrm{d}\mathbf{x}\,\mathrm{d}t = 0. \tag{5.48}$$

□

The asymptotic error committed by using the Taylor expansion to approximate the moments, is stated in the following, see [16, 30].

Proposition 5.2 *By means of the stochastic Taylor expansion (5.46), with* $\mathrm{E}[\tilde{\nu}_C] = 0$, *the expected value and the variance can be expanded as*

$$\mathrm{E}[\hat{Q}_s] = \hat{Q}_0 + \mathcal{O}(s^2), \tag{5.49}$$

$$\mathrm{Var}[\hat{Q}_s] = s^2\mathrm{E}[\delta\hat{Q}^2] + \mathcal{O}(s^3). \tag{5.50}$$

Moreover, for the k-th central moment we have

$$\mathrm{M}^k[\hat{Q}_s] = s^k\mathrm{E}[\delta\hat{Q}^k] + \mathcal{O}(s^{k+1}). \tag{5.51}$$

Proof We immediately deduce from Lemma 5.1 that

$$\mathrm{E}[\hat{Q}_s] = \mathrm{E}[\hat{Q}_0 + s\delta\hat{Q} + \mathcal{O}(s^2)] = \hat{Q}_0 + \mathcal{O}(s^2). \tag{5.52}$$

Generalization to the k-th statistical moment yields

$$\mathrm{M}^k[\hat{Q}] = s^k\mathrm{E}[\delta\hat{Q}^k] + \mathcal{O}(s^{k+1}), \tag{5.53}$$

as can be seen by

$$\mathrm{M}^k[\hat{Q}_s] = \mathrm{E}[(s\delta\hat{Q} + \mathcal{O}(s^2))^k] \tag{5.54}$$
$$= s^k \mathrm{E}[\delta\hat{Q}^k] + \mathcal{O}(s^{k+1}), \tag{5.55}$$

cf. [27, p. 14]. The result for the variance follows for $k = 2$. □

By the previous result we compute approximations of the moments as

$$\mathrm{E}_s[\hat{Q}] = \hat{Q}_0, \tag{5.56}$$
$$\mathrm{Var}_s[\hat{Q}] = s^2 \mathrm{E}[\delta\hat{Q}^2], \tag{5.57}$$
$$\mathrm{M}_s^k[\hat{Q}] = s^k \mathrm{E}[\delta\hat{Q}^k], \tag{5.58}$$

with accuracy of order two, three and $k + 1$ for the mean, variance and k-th moment, respectively. Their evaluation requires the solution of the state and adjoint equation at the nominal value ν_{C} as well as the computation of the high-dimensional integral $\mathrm{E}[\delta\hat{Q}^k]$. Provided that high-dimensional integration can be carried out at moderate cost, the overall cost of the perturbation method can be considered small with respect to classical Monte Carlo simulation. Indeed, here, only the state and adjoint equation have to be solved as opposed to solving the state problem hundred or thousand times. Hence, in a next step, we address the evaluation of $\mathrm{E}[\delta\hat{Q}^k]$. To simplify notation we set $\mathbf{s} := (t, \mathbf{x})$, $S := I_T \times D_{\mathrm{C}}$ and $\alpha(\mathbf{s}) := \mathbf{grad}\ u(\mathbf{s}) \cdot \mathbf{grad}\ p(\mathbf{s})$, where u and p refer to the nominal solution of state and adjoint equation, respectively. We obtain

$$\mathrm{E}[\delta\hat{Q}^k] = \int_\Omega \left(\int_S \tilde{\nu}_{\mathrm{C}}(\omega, B(\mathbf{s}))\alpha(\mathbf{s})\ \mathrm{d}\mathbf{s} \right)^k \mathrm{dP} = \int_{S^{\times k}} \underbrace{\left(\int_\Omega \tilde{\nu}(\omega, B(\mathbf{s}_1)) \cdots \tilde{\nu}(\omega, B(\mathbf{s}_k))\ \mathrm{dP} \right)}_{=:\mathrm{Cor}^k_{\tilde{\nu}_{\mathrm{C}}}(B(\mathbf{s}_1),\ldots,B(\mathbf{s}_k))} \alpha(\mathbf{s}_1) \cdots \alpha(\mathbf{s}_k)\ \mathrm{d}\mathbf{s}_1 \cdots \mathrm{d}\mathbf{s}_k, \tag{5.59}$$

where $S^{\times k}$ refers to the k-fold Cartesian product. Solving state and adjoint equation by the lowest order finite element method and the backward Euler time-stepping procedure yields approximations u_{h,h_T}, p_{h,h_T} and B_{h,h_T}, α_{h,h_T}, respectively. For simplicity, the linearization error is omitted here. Using these approximations in (5.59) yields

$$\mathrm{E}[(\delta\hat{Q}_{h,h_T})^k] = \int_{S^{\times k}} \underbrace{\mathrm{Cor}^k_{\tilde{\nu}}(B_{h,h_T}(\mathbf{s}_1), \ldots, B_{h,h_T}(\mathbf{s}_k))\alpha_{h,h_T}(\mathbf{s}_1) \cdots \alpha_{h,h_T}(\mathbf{s}_k)}_{=:\gamma^{\times k}_{h,h_T}(\mathbf{s}_1,\ldots,\mathbf{s}_k)}\ \mathrm{d}\mathbf{s}_1 \cdots \mathrm{d}\mathbf{s}_k. \tag{5.60}$$

Due to lowest order discretization schemes the integrand $\gamma_{h,h_T}^{\times k}$ in the previous equation is constant over each space-time element $S_n := I_{i(n)} \times K_{j(n)}$, where $I_i \in \mathcal{T}_h$ and $K_j \in \mathcal{T}_T$, respectively. Hence, we obtain

$$\mathrm{E}[(\delta \hat{Q}_{h,h_T})^k] = \sum_{i_1=1}^{N_h N_T} \cdots \sum_{i_k=1}^{N_h N_T} |S_{i_1}| \dots |S_{i_k}| (\gamma_{h,h_T})^{\times k} (\tilde{\mathbf{s}}_1^{i_1}, \dots, \tilde{\mathbf{s}}_k^{i_k}), \tag{5.61}$$

where $\tilde{\mathbf{s}}_1^{i_j}$ is an arbitrary point in S_{i_j} and $|S_i| = \mathrm{meas}(S_i)$. Although, for a moderate size of $N_h N_T$ and k, the effort of directly computing (5.61) is affordable, the cost grows with $\mathcal{O}((N_h N_T)^k)$. Hence, for higher order moments and fine discretizations alternative quadrature techniques are required. In this context sparse grid constructions in combination with multilevel schemes were proposed, see, e.g., [27, 30]. Then under certain assumptions on the smoothness of the integrand the growth of the cost can be limited to essentially $\mathcal{O}(N_h N_T)$. Here we propose to use Monte Carlo integration techniques to cope with the issue of high-dimensional integration because of its simplicity. The procedure is as follows: we define a rectangular domain $\hat{S} \subset \mathbb{R}^3$ such that $I_T \times D_C \subset \hat{S}$ holds. Next, we generate N^{MC} pseudo-random numbers $\mathbf{r}^i$, $i = 1, \dots, N^{\mathrm{MC}}$, in $\hat{S}^{\times k}$ with an underlying uniform distribution. Then (5.61) is approximated by means of

$$\mathrm{E}_{\mathrm{MC}}[(\delta \hat{Q}_{h,h_T})^k] \approx \frac{|I_T \times D_C|}{N^{\mathrm{MC}}} \sum_{i=1}^{N^{\mathrm{MC}}} \gamma_{h,h_T}^{\times k}(\mathbf{r}^i) \chi_{I_T \times D_C}(\mathbf{r}^i). \tag{5.62}$$

The (probabilistic) integration error can be controlled as argued in Sect. 5.2.2. The price to pay for this simple and dimension-independent procedure is the slow convergence rate with respect to N^{MC}.

To complete the numerical scheme, the k-point correlation function $\mathrm{Cor}_{\tilde{\nu}}^k$ needs to be discretized. For simplicity, we consider the case $k = 2$, a generalization is straight forward. As in the present adjoint setting the cost is independent of the number of inputs, we can directly choose the spline model (4.9). Perturbing the points H_i^{ms} randomly as $H_i^{\mathrm{ms}}(\omega) = H_i^{\mathrm{ms}} + s\tilde{H}_i^{\mathrm{ms}}(\omega)$, we obtain a perturbed spline

$$f_{HB}(\omega, B) = f_{HB}(B) + s\tilde{\mathbf{H}}^{\mathrm{ms}}(\omega) \cdot \boldsymbol{\phi}(B), \tag{5.63}$$

where $\boldsymbol{\phi}$ contains cubic Hermite basis function as outlined in Sect. 4.2.2. The reluctivity is then defined as $\nu(\omega, B) = f_{HB}(\omega, B)/B$. Since the $\tilde{H}_i^{\mathrm{ms}}$ are centered, in this finite dimensional setting, the correlation function can be expressed through the covariance matrix $\mathbf{V}$ by

$$\mathrm{Cor}_{\tilde{\nu}}(B_1, B_2) = \hat{\boldsymbol{\phi}}^\top(B_1) \mathbf{V}_{\tilde{\mathbf{H}}^{\mathrm{ms}}} \hat{\boldsymbol{\phi}}(B_2), \tag{5.64}$$

where $\hat{\boldsymbol{\phi}}(B) = \boldsymbol{\phi}(B)/B$. For independent and uniformly distributed $\tilde{H}_i^{\mathrm{ms}}$, i.e., $\tilde{H}_i^{\mathrm{ms}} \sim \mathcal{U}(-1, 1)$, we obtain $\mathbf{V}_{\tilde{\mathbf{H}}^{\mathrm{ms}}} = 1/3\mathbf{I}$.

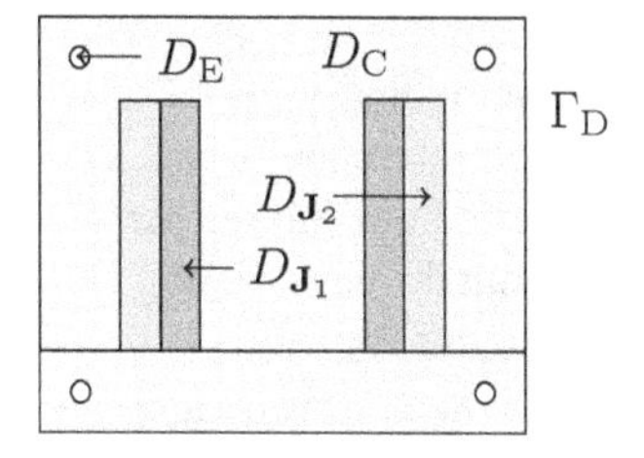

Fig. 5.3 Figure after [19]. Model geometry of a transformer [31]. Primary and secondary coils are denoted $D_{\mathbf{J},1}$ and $D_{\mathbf{J}_2}$, respectively

The key benefit of the adjoint approach is the independence with respect to the number of input variables. However, for multiple quantities of interest, the adjoint problem has to be solved repetitively. If the number of inputs, i.e., M is small with respect to the number of QoIs, a finite difference approach might be favorable, especially as it is non-intrusive. In this case, for each input Y_i, the derivative is approximated by the finite difference quotient

$$\frac{\partial \hat{Q}}{\partial Y_i} \approx \frac{\hat{Q}(\mathbf{Y} + s\tilde{Y}_i \mathrm{e}_i) - \hat{Q}(\mathbf{Y})}{s} \tag{5.65}$$

and $s > 0$. This in turn would require the repetitive solution of the state problem, see [17] for a more detailed description.

Example 5.2 From [19] we report a simple two-dimensional example here, see Fig. 5.3 for the geometry. For details on the geometry and the material data see [31]. Both conductors $D_{\mathbf{J},1}$ and $D_{\mathbf{J},2}$ are modeled as stranded conductors, i.e., the conductivity is set to zero, whereas the iron conductivity of the solid iron is 2.9MS. An excitation current is imposed to the primary coil $D_{\mathbf{J},1}$ and the secondary coil is left open. Both transient and static simulations are carried out, referred to as parabolic (A) and elliptic (B) case, respectively. The imposed currents are $I = 0.012$A (A) and $I(t) = 0.12\sin(\pi t/T)$A (B), with $T = 0.02$s, respectively. Discretization is carried out by lowest order nodal finite elements in space with 4571 nodes using FEMM [31] and the implicit Euler method with a time-step of $h_T = T/90$. The in-house MATLAB code NIOBE is used for the simulation. Finally, Newton's method is employed for linearization based on Algorithm A.1 and the error is controlled by means of (3.71) with a relative tolerance of 1×10^{-6}.

We recall from [32] an expression for the inductance at time t_i as

$$L(t_i) = \mathbf{P}_{\mathrm{str}}^T \mathbf{K}(\mathbf{u}_i)^{-1} \mathbf{P}_{\mathrm{str}} = \frac{\mathbf{P}_{\mathrm{str}}^T \mathbf{u}_i}{I(t_i)}, \tag{5.66}$$

for $I(t_i) \neq 0$, with stiffness matrix $\mathbf{K}$ and $\mathbf{P}_{\mathrm{str}}$ the stranded conductor coupling vector, see, e.g., [33]. For the cases (A) and (B) the average value of $L(t)$ over T and $\mathbf{P}_{\mathrm{str}}^T \mathbf{u}/I$, are chosen as QoI, respectively. Random data for the given material [31] is obtained by introducing uncorrelated perturbations $\tilde{H}_i^{\mathrm{ms}}$ with a maximum amplitude of approximately 6 %. As a reference, in Example 5.1 perturbation amplitudes between 2–6 % were considered. The perturbation method is compared to the Monte Carlo method with Latin hypercube sampling. To reduce the computational effort for the reference Monte Carlo simulation, a sensitivity analysis for each point H_i^{ms} is carried out and the perturbations are restricted to the three most influential values for both the perturbation and the Monte Carlo method. The difference in the variance due to this reduction is found to be below 6 %. Random perturbations $\tilde{H}_i^{\text{ms}}$ are modeled to be uniformly distributed as $\tilde{H}_i \sim sH_i\mathcal{U}(-1, 1)$. In general, the error indicator, defined as the difference between the Monte Carlo simulation and the perturbation method is denoted Δ_{MC}.

For the static, elliptic case (A), in Fig. 5.4, Δ_{MC} for mean and variance obtained by 10000 Monte Carlo samples, is depicted for different magnitudes of perturbation s. As the corresponding slopes are 1.98 and 3.69, respectively, the asymptotic convergence rates with respect to s given in Proposition 5.2 are confirmed. Note that for the variance, the convergence is even higher than predicted. Concrete values of mean variance and error indicators are partially given in Table 5.1, whereas results for the time-transient case (A) are reported in Table 5.2. In the latter case, only 1000 samples were used for the Monte Carlo reference as the computational complexity is significantly increased. Consequently, the asymptotic convergence with respect to s cannot be observed, however, we obtain an estimate of the error magnitude. In both cases the estimated error level is comparable and in the same order of magnitude as the input uncertainty. In conclusion for this example, the perturbation method

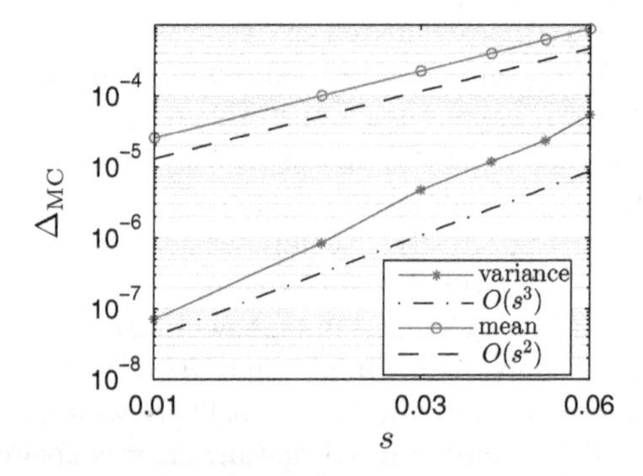

Fig. 5.4 Figure after [19]. Estimated errors of the perturbation method, by means of a Monte Carlo simulation with 10000 samples. The convergence rates are 1.98 and 3.69 for mean value and variance, respectively

Table 5.1 Results from [19]. Estimated errors of the perturbation method, by means of a Monte Carlo simulation with 10000 samples for the elliptic case (B)

s	$\text{E}_s[L]/\text{H}$	$\Delta_{\text{MC}}\%$	$\text{Var}_s[L]/\text{H}$	$\Delta_{\text{MC}}\%$
0.06	3.576	0.024	1.02×10^{-3}	5.6
0.03	3.576	0.006	2.54×10^{-4}	1.9
0.01	3.576	0.001	2.82×10^{-5}	0.2

Table 5.2 Results from [19]. Estimated errors of the perturbation method, by means of a Monte Carlo simulation with 1000 samples for the parabolic case (A)

s	$\mathrm{E}_s[L]/\mathrm{H}$	$\Delta_{\mathrm{MC}}\%$	$\mathrm{Var}_s[L]/\mathrm{H}$	$\Delta_{\mathrm{MC}}\%$
0.06	3.616	0.023	2.07×10^{-4}	4.4
0.03	3.616	0.006	5.18×10^{-5}	4.0
0.01	3.616	0.001	5.76×10^{-6}	3.8

provides estimates with moderate errors at significantly reduced cost, with respect to the classical Monte Carlo method.

5.2.4 Collocation Based Polynomial Chaos Method

Stochastic collocation and Galerkin methods, also referred to as stochastic (pseudo)-spectral methods, have received considerable attention in recent years [5, 11, 34–36]. By exploiting the possible high regularity of the solution in the stochastic parameter space Γ, spectral convergence is obtained in many situations. Hence, the computational cost can be significantly reduced with respect to classical Monte Carlo methods for moderate input dimensions. Moreover, sparse-grid constructions have been proposed as a means to tackle models with a large number of uncertain inputs. Here, we are in particular interested in the stochastic collocation method. It is non-intrusive, i.e., only repetitive solutions of the original deterministic problem are required and the application to nonlinear time-dependent problems is straight forward. For a comparison of stochastic collocation and Galerkin methods in the context of the simulation of electrical machines by means of a nonlinear convection-diffusion problem we refer to [37].

A key ingredient for the stochastic collocation and Galerkin method is a global polynomial Ansatz for the QoI $\hat{Q} : \Gamma \to \mathbb{R}$, or any other square-integrable function. For instance, the generalized Polynomial Chaos (gPC) expansion reads as

$$\hat{Q} = \sum_{i=0}^{\infty} a_i \Xi_i(Y_1, \ldots, Y_n). \tag{5.67}$$

In (5.67) the Ξ_i are polynomials, orthogonal with respect to the probability distribution of Y_i, i.e.,

$$\mathrm{E}[\Xi_i \Xi_j] = \begin{cases} \mathrm{E}[\Xi_i^2], & i = j, \\ 0, & i \neq j. \end{cases} \tag{5.68}$$

In this respect, Hermite and Legendre polynomials are associated to normally and uniformly distributed random variables, respectively. We refer to [5] for other types of distributions. Omitting polynomials of (total) degree greater than q in (5.67), an expansion suitable for computational purposes is obtained. Once this surrogate model is at hand, statistical moments, probability distributions and global sensitivities, e.g., coefficients in an ANOVA-like expansion such as (5.34), are obtained at a low cost.

Here, we describe the approximation by the stochastic collocation method, based on Lagrange polynomials [5, Chap. 7]. Starting point for the discussion is (5.33), i.e.,

$$\int_{D_C} \sigma_C \dot{\mathbf{A}}(\mathbf{y}, t, \cdot) \cdot \mathbf{v}\, \mathrm{d}\mathbf{x} + \int_D \nu(\mathbf{y}, \cdot, |\mathbf{curl}\, \mathbf{A}(\mathbf{y}, t, \cdot)|) \mathbf{curl}\, \mathbf{A}(\mathbf{y}, t, \cdot) \cdot \mathbf{curl}\, \mathbf{v}\, \mathrm{d}\mathbf{x} = \int_D \mathbf{J}(t) \cdot \mathbf{v} dx, \tag{5.69}$$

for (ϱ-) almost all $\mathbf{y} \in \Gamma$. For simplicity, in (5.69) and in the remaining part of this subsection, we consider a variable reluctivity, solely. Following [3], consider a tensor product grid

$$(\mathbf{y}_r)_{r=1}^{N^q} = \{y_1^1, y_1^2, \ldots, y_1^{m_1}\} \times \{y_2^1, y_2^2, \ldots, y_2^{m_2}\} \times \cdots \times \{y_M^1, y_M^2, \ldots, y_M^{m_M}\}, \tag{5.70}$$

where in each dimension $n = 1, \ldots, M$, we have $m_n = q_n + 1$ collocation points and $N^q = m_1 m_2 \ldots m_M$. Collocation points are chosen in each direction as the roots of the orthogonal polynomials associated to the probability density function ϱ_i. In particular for a uniform distribution we choose the roots of Legendre polynomials as stated above. An extension to sparse grid constructions is given in [3, 38, 39]. To a set of local indices $(r_1, r_2, \ldots, r_M)$, where $1 \leq r_n \leq q_n + 1$, the global index r is associated by

$$r = r_1 + q_1(r_2 - 1) + q_1 q_2(r_3 - 1) + \cdots. \tag{5.71}$$

Following the one-dimensional approach of [3, 15], we set $\mathbf{y} = (y_n, \hat{\mathbf{y}}_n)$, $\hat{\mathbf{y}}_n = (y_1, \ldots, y_{n-1}, y_{n+1}, \ldots, y_M)$. Let $\mathbb{Q}_{q_n}$ denote the space of polynomials of degree at most q_n and $\mathbb{Q}_q$ the tensor product space of polynomials of degree at most q_n in each direction, respectively. The one-dimensional Lagrange interpolation operator $\mathcal{I}_{q_n} : C(\Gamma, W(D)) \to \mathbb{Q}_{q_n}(\Gamma_n) \otimes W(D)$ is defined as

$$\mathcal{I}_{q_n}\mathbf{u}(\mathbf{y}) = \sum_{i=1}^{m_n} \mathbf{u}(y_n^i, \hat{\mathbf{y}}_n) l_n^i(y_n), \tag{5.72}$$

where $l_n^i(y_n)$ is the Lagrange polynomial of degree $q_n - 1$ associated to the point y_n^i. A tensor product interpolation formula is readily obtained as

$$\mathcal{I}_{\mathbf{q}}\mathbf{u}(\mathbf{y}) = \mathcal{I}_{q_1} \otimes \cdots \otimes \mathcal{I}_{q_M}\mathbf{u}(\mathbf{y}) = \sum_{r=1}^{N^q} \mathbf{u}(\mathbf{y}_r) l_r(\mathbf{y}), \tag{5.73}$$

where $l_r(\mathbf{y})$ denotes the global Lagrange polynomial associated to the point $\mathbf{y}_r$. By solving (5.69) for all $\mathbf{y}_r$, $r = 1, \ldots, N^q$ and interpolating the solution as

$$\mathbf{A}_q(\mathbf{y}) = \sum_{r=1}^{N^q} \mathbf{A}(\mathbf{y}_r) l_r(\mathbf{y}) \tag{5.74}$$

we complete the collocation procedure. Finally, the expected value is approximated by the quadrature

$$\mathrm{E}_q[\hat{Q}] = \sum_{r=1}^{N^q} w_r \hat{Q}(\mathbf{y}_r), \tag{5.75}$$

where the weights are given by $w_r = \int_\Gamma l_r^2(\mathbf{y}) \varrho(\mathbf{y})\, \mathrm{d}\mathbf{y}$ and approximations of other statistical moments can be obtained in the same way.

Convergence estimates of the collocation method depend on the regularity of the mapping

$$\mathbf{y} \mapsto \mathbf{A}(\mathbf{y}). \tag{5.76}$$

It is well-known, that for linear elliptic problems this mapping is analytic [3, 4, 40] which results in an exponential convergence of stochastic collocation and Galerkin schemes. Here, in view of the nonlinear reluctivity, the situation is slightly different as it has been observed [41, 42], that the magnetoquasistatic model (5.76) fails to be complex differentiable. Still, higher order differentiability can be established as outlined in the following. We consider the static case with a single input, i.e., $M = 1$. We have covered the case of general M in [6]. Following [3], differentiating

$$(\nu(y, \cdot, |\mathbf{curl}\ \mathbf{A}(y)|)\mathbf{curl}\ \mathbf{A}(y), \mathbf{curl}\ \mathbf{v})_D = (\mathbf{J}, \mathbf{v})_D, \quad \forall \mathbf{v} \in W_{\mathrm{st}}(D)$$

k-times with respect to y we can characterize $\partial_y^k \mathbf{A}$ by the boundary value problem

$$\left(\nu_{\mathrm{d}}(y, \mathbf{curl}\ \mathbf{A})\mathbf{curl}\ \partial_y^k \mathbf{A}, \mathbf{curl}\ \mathbf{v}\right)_D = -\left(\mathbf{G}_k, \mathbf{curl}\ \mathbf{v}\right)_D, \tag{5.77}$$

for all $\mathbf{v} \in W_{\mathrm{st},h}(D)$. More precisely, as shown in [6] or Appendix C, we have

$$\mathbf{G}_k := \sum_{l=0}^{k} \binom{k}{l} \sum_{\substack{\pi \in \Pi_{k-l}, \\ |\pi| \neq 1, l=0}} \partial_y^l D_{\mathbf{r}}^{|\pi|} \mathbf{h}(y, \mathbf{curl}\ \mathbf{A}) \left(\partial_y^{|\pi_1|} \mathbf{curl}\ \mathbf{A}, \ldots, \partial_y^{|\pi_{|\pi|}|} \mathbf{curl}\ \mathbf{A}\right), \tag{5.78}$$

where Π_k is the set of all partitions of $\{1, 2, \ldots, k\}$ and $|\pi|$ the number of blocks in $\pi = \{\pi_1, \ldots, \pi_{|\pi|}\}$. Also $D^2\mathbf{h}$ and $D^k\mathbf{h}$ refer to the Hessian matrix and higher order derivatives, respectively, subject to the following assumption:

Assumption 5.2 For all $l = 1, \ldots, k$ and $\partial_y^{k-l} D^l \mathbf{h}(y, \cdot)$ there holds

$$|\partial_y^{k-l} D^l \mathbf{h}(y, \mathbf{r})(\mathbf{r}_1, \ldots, \mathbf{r}_k)| \leq C_k |\mathbf{r}_1| \cdots |\mathbf{r}_k|. \tag{5.79}$$

For a discussion of this assumption, see [6]. The aim of parametric regularity analysis is to characterize $\partial_y^k \mathbf{A}$ as an element of $W_{\text{st}}(D)$. To this end we have to show that $\mathbf{G}_k \in L^2(D)^3$, which in turn follows from

$$\| \, |\partial_y^{|\pi_1|} \mathbf{curl}\, \mathbf{A}| \cdots |\partial_y^{|\pi_{|\pi|}|} \mathbf{curl}\, \mathbf{A}| \, \|_2 < \infty, \tag{5.80}$$

by the continuity of $D^k \mathbf{h}$. This is the basis for the following algebraic convergence estimate of the stochastic error, where for the proof we again refer to Appendix C or [6].

Theorem 5.1 *Let $\nu_{\text{d}}(\mathbf{y}, \cdot)$ fulfill Assumption 3.10 (ϱ-) a.e. and Assumption 5.2 hold true. The collocation approximation $\mathbf{A}_q$ converges to $\mathbf{A}$, as*

$$\|\mathbf{A} - \mathbf{A}_q\|_{L^2_\varrho(\Gamma)\otimes W_{\text{st}}(D)} \leq C_1 q^{-1}. \tag{5.81}$$

Additionally, the collocation error for the finite element solution $\mathbf{A}_h$ converges as

$$\|\mathbf{A}_h - \mathbf{A}_{h,q}\|_{L^2_\varrho(\Gamma)\otimes W_{\text{st}}(D)} \leq C_2 q^{-k}, \tag{5.82}$$

where C_2 depends on k, h.

Collocation can be carried out after finite element approximation, as we observe that both commute. We also note that the estimate (5.82) in Theorem 5.1 deteriorates in the limit $h \to 0$, as $C_2 \to \infty$ for $h \to 0$. In [6] we also consider the finite element discretization and the linearization error.

As the total number of points of the tensor grid is $N^q = (1+q)^M$, the previous result states that, asymptotically, $(N^q)^{-k/M}$ evaluations are needed for a specified stochastic error level. Comparing with the Monte Carlo rate of $(N^{\text{MC}})^{-1/2}$ we expect the collocation method to be asymptotically superior for $k/M > 1/2$, see [38] for a more detailed cost analysis. This holds true in particular for high parametric regularity and a moderate number of input parameters.

Remark 5.2 In Theorem 5.1 estimate (5.81) can be improved to q^{-k} for $k > 1$, using (5.80) and the Hölder inequality, provided a spatial regularity result such as $|\mathbf{curl}\; \partial_y^l \mathbf{A}| \in L^{2k}(D), l < k$, can be established. For a smooth domain and data, (5.80) would be bounded for all $k \in \mathbb{N}$.

5.2.5 Worst-Case Scenario

Techniques from optimization are also used in the context of (epistemic) uncertainty quantification, referred to as Worst-Case Scenario (WCS), for various reasons. An accurate characterization of the probability distribution of uncertain inputs may require a large amount of experiments which might not always be affordable or even unachievable. Also, when the production of a device is costly or its failure would be critical for a specific application, computing a worst-case rather than some moments would be desirable. Moreover, using adjoint based optimization methods can be favorable for high-dimensional problems and also in the context of robust optimization, where appropriate routines might already be at hand. Following [43, 44] a worst-case scenario consists in computing

$$\mathrm{wcs}(\hat{Q}) := \sup_{\boldsymbol{\beta} \in U_{\mathrm{adm}}} |\hat{Q}[\boldsymbol{\beta}] - \hat{Q}[\boldsymbol{\beta}_0]|, \tag{5.83}$$

where $\boldsymbol{\beta}_0$ refers to a nominal value. Using the theory of [44, p. 52], the existence of a solution of (5.83), denoted $\boldsymbol{\beta}^+$, can be established under some mild assumptions on U_{adm}, using the unique solvability of the magnetoquasistatic model and the linearity of the QoI. Techniques from optimization can be applied directly to (5.83). Here, following [43], we apply again a perturbation technique, aiming at a reduced computational cost. By means of the Fréchet derivative $< \delta\hat{Q}, \tilde{\boldsymbol{\beta}} >$ (where we explicitly indicate the perturbation $\tilde{\boldsymbol{\beta}}$ direction and still omit the evaluation point $\boldsymbol{\beta}_0$) Taylor's expansion reads as

$$\hat{Q}_s = \hat{Q}_0 + s < \delta\hat{Q}, \tilde{\boldsymbol{\beta}} > + s^2 \frac{1}{2} D^2 \hat{Q}[\boldsymbol{\beta}_0 + \theta\tilde{\boldsymbol{\beta}}](\tilde{\boldsymbol{\beta}}, \tilde{\boldsymbol{\beta}}), \tag{5.84}$$

where $\theta \in (0, 1)$ and D^2 refers to the second derivative, cf. [43]. Using (5.84) in (5.83) we obtain

$$\mathrm{wcs}(\hat{Q}) \leq \underbrace{s \sup_{\tilde{\boldsymbol{\beta}}} | < \delta\hat{Q}, \tilde{\boldsymbol{\beta}} > |}_{=:\mathrm{wcs}_{\mathrm{L}}(\hat{Q})} + \underbrace{s^2 \frac{1}{2} \sup_{\tilde{\boldsymbol{\beta}}} \sup_{\theta \in (0,1)} |D^2 \hat{Q}[\boldsymbol{\beta}_0 + \theta\tilde{\boldsymbol{\beta}}](\tilde{\boldsymbol{\beta}}, \tilde{\boldsymbol{\beta}})|}_{\text{remainder}}, \tag{5.85}$$

see [43, p. 6], where the supremum is sought over all $\tilde{\boldsymbol{\beta}}$ such that $\boldsymbol{\beta}_s(\tilde{\boldsymbol{\beta}}) \in U_{\mathrm{adm}}$. Neglecting the remainder in the previous expression, we are now concerned with a linear, though infinite-dimensional problem. It has been shown [43], for a specific form of the set of admissibility and perturbations in L^∞, that $\mathrm{wcs}_{\mathrm{L}}(\hat{Q})$ can be computed exactly. Unfortunately, this is not readily extended to the inputs under consideration in this work, mainly due to our smoothness requirements. However, the situation is different in the case of finite-dimensional inputs. Then the WCS reads as

$$\mathrm{wcs}_{\mathrm{L}}(\hat{Q}) = \sup_{\delta\mathbf{y} \in \Gamma - \mathbf{y}_0} |\mathbf{grad}_{\mathbf{y}} \hat{Q}[\boldsymbol{\beta}(\mathbf{y})] \cdot \delta\mathbf{y}|, \tag{5.86}$$

which is a linear programming problem with box constraints. Let $|y_i - y_{0,i}| \leq |\Gamma_i|$, the solution and the worst-case are given as

$$\mathrm{wcs}_{\mathrm{L}}(\hat{Q}) = \sum_{i=1}^{M} |\partial_{y_i} \hat{Q}[\boldsymbol{\beta}(\mathbf{y})]||\Gamma_i| \tag{5.87}$$

$$y_i^+ = y_{i,0} + \mathrm{sign}(\partial_{y_i} \hat{Q}[\boldsymbol{\beta}(\mathbf{y})])|\Gamma_i|, \tag{5.88}$$

respectively, cf. [43, p. 9]. Finally, let us comment on uncertainty modeling in this case. Box constraints might be appropriate for an underlying uniform random distribution as no preference for points around the nominal value is imposed. In contrary, for normally distributed random vectors, realizations at the corners of Γ are very unlikely and should be excluded to enhance the accuracy of the WCS prediction. In this case, given the covariance matrix $\mathbf{V}_\mathbf{y}$ of the inputs, we set

$$\Gamma = \{\mathbf{y} \mid (\mathbf{y} - \mathbf{y}_0)^\top \mathbf{V}_\mathbf{y}^{-1} (\mathbf{y} - \mathbf{y}_0) \leq \theta^2\}, \tag{5.89}$$

where θ is a safety parameter near 1 [45].

5.3 Conclusion

In conclusion, in this section, we have discussed methods for uncertainty quantification, in particular concerning uncertainty modeling and propagation. A stochastic formulation of the magnetoquasistatic model was stated together with a high-dimensional deterministic counterpart. Discretization of the random inputs was achieved by the Karhunen-Loève expansion. After identifying high-dimensionality as a key problem in uncertainty quantification, a method for dimension reduction, the cut-HDMR expansion, was discussed. Then, in the context of uncertainty propagation, several methods were extended to the present setting. Adjoint techniques within moment based methods and worst-case scenarios are appealing to cope with a large number of inputs, whereas typically code modifications are required and error control is not fully established in general. In contrary, non-intrusive methods are set up easily. However, for the Monte Carlo method, the convergence rate is limited, whereas for the (tensor grid) gPC based collocation method, the curse of dimensionality limits the application to small up to moderate M. As illustrated in Example 5.2, cheap perturbation techniques, approximating only the linear effects in the HDMR expansion, might be used even for the present nonlinear model.

References

1. Loève, M.: Probability Theory. Foundations. Random Sequences. D. Van Nostrand Company, New York (1955)
2. Adler, R.J.: The Geometry of Random Fields. SIAM (1981)
3. Babuška, I., Nobile, F., Tempone, R.: A stochastic collocation method for elliptic partial differential equations with random input data. SIAM Rev. **52**(2), 317–355 (2010)
4. Charrier, J.: Strong and weak error estimates for elliptic partial differential equations with random coefficients. SIAM J. Numer. Anal. **50**(1), 216–246 (2012)
5. Xiu, D.: Numerical Methods for Stochastic Computations: A Spectral Method Approach. Princeton University Press (2010)
6. Römer, U., Schöps, S., Weiland, T.: Stochastic modeling and regularity of the nonlinear elliptic curl-curl equation. SIAM/ASA J Uncertainty Quantification (in press)

7. Schwab, C., Todor, R.A.: Karhunen-Loève approximation of random fields by generalized fast multipole methods. J. Comput. Phys. **217**(1), 100–122 (2006)
8. Loève, M.: Probability theory. Grad. Texts Math. **45**, 12 (1963)
9. Frauenfelder, P., Schwab, C., Todor, R.A.: Finite elements for elliptic problems with stochastic coefficients. Comput. Methods Appl. Mech. Eng. **194**(2), 205–228 (2005)
10. Babuška, I., Liu, K.-M., Tempone, R.: Solving stochastic partial differential equations based on the experimental data. Math. Models Methods Appl. Sci. **13**(03), 415–444 (2003)
11. Babuška, I., Tempone, R., Zouraris, G.E.: Galerkin finite element approximations of stochastic elliptic partial differential equations. SIAM J. Numer. Anal. **42**(2), 800–825 (2004)
12. De Boor, C.: A Practical Guide to Splines, vol. 27. Springer, New York (1978)
13. Ramarotafika, R., Benabou, A., Clénet, S.: Stochastic modeling of soft magnetic properties of electrical steels, application to stators of electrical machines. IEEE Trans. Magn. **48**, 2573–2584 (2012)
14. Deb, M.K., Babuška, I.M., Oden, J.T.: Solution of stochastic partial differential equations using galerkin finite element techniques. Comput. Methods Appl. Mech. Eng. **190**(48), 6359–6372 (2001)
15. Nobile, F., Tempone, R.: Analysis and implementation issues for the numerical approximation of parabolic equations with random coefficients. Int. J. Numer. Methods Eng. **80**(6–7), 979–1006 (2009)
16. Harbrecht, H., Schneider, R., Schwab, C.: Sparse second moment analysis for elliptic problems in stochastic domains. Numer. Math. **109**(3), 385–414 (2008)
17. Harbrecht, H., Li, J.: First order second moment analysis for stochastic interface problems based on low-rank approximation. ESAIM: Math. Model. Numer. Anal. **47**(05), 1533–1552 (2013)
18. Stiemer, M.: A Galerkin method for mixed parabolic-elliptic partial differential equations. Numer. Math. **116**(3), 435–462 (2010)
19. Römer, U., Schöps, S., Weiland, T.: Approximation of moments for the nonlinear magneto-quasistatic problem with material uncertainties. IEEE Trans. Magn. **50**(2) (2014)
20. Labovsky, A., Gunzburger, M.: An efficient and accurate method for the identification of the most influential random parameters appearing in the input data for PDEs. SIAM/ASA J. Uncertain. Quantif. **2**(1), 82–105 (2014)
21. Sobol, I.M.: On sensitivity estimation for nonlinear mathematical models. Matematicheskoe Modelirovanie **2**(1), 112–118 (1990)
22. Dick, J., Kuo, F.Y., Sloan, I.H.: High-dimensional integration: the quasi-Monte Carlo way. Acta Numerica **22**, 133–288 (2013)
23. Cliffe, K.A., Giles, M.B., Scheichl, R., Teckentrup, A.L.: Multilevel Monte Carlo methods and applications to elliptic PDEs with random coefficients. Comput. Vis. Sci. **14**(1), 3–15 (2011)
24. Barth, A., Schwab, C., Zollinger, N.: Multi-level Monte Carlo finite element method for elliptic PDEs with stochastic coefficients. Numer. Math. **119**(1), 123–161 (2011)
25. Schwab, Ch., Todor, R.-A.: Sparse finite elements for stochastic elliptic problems-higher order moments. Computing **71**(1), 43–63 (2003)
26. Schwab, C., Todor, R.-A.: Sparse finite elements for elliptic problems with stochastic loading. Numer. Math. **95**(4), 707–734 (2003)
27. Chernov, A., Schwab, C.: First order k-th moment finite element analysis of nonlinear operator equations with stochastic data. Math. Comput. **82**(284), 1859–1888 (2013)
28. Harbrecht, H., Peters, M., Siebenmorgen, M.: Combination technique based k-th moment analysis of elliptic problems with random diffusion. J. Comput. Phys. **252**, 128–141 (2013)
29. Bonizzoni, F., Buffa, A., Nobile, F.: Moment equations for the mixed formulation of the hodge laplacian with stochastic loading term. IMA J. Numer. Anal. (2013)
30. Harbrecht, H.: On output functionals of boundary value problems on stochastic domains. Math. Methods Appl. Sci. **33**(1), 91–102 (2010)
31. Meeker, D.: Finite element method magnetics. Version 4.2 (1 April 2009 Build) (2010)
32. Schöps, S.: Multiscale modeling and multirate time-integration of field/circuit coupled problems. Ph.D. thesis, Katholieke Universiteit Leuven (2011)

33. Schöps, S., De Gersem, H., Weiland, T.: Winding functions in transient magnetoquasistatic field-circuit coupled simulations. COMPEL: Int. J. Comput. Math. Electr. Electron. Eng. **32**(6), 2063–2083 (2013)
34. Ghanem, R.G., Spanos, P.D.: Stochastic Finite Elements: a Spectral Approach. Springer (1991)
35. Xiu, D., Karniadakis, G.E.: The Wiener-Askey polynomial chaos for stochastic differential equations. SIAM J. Sci. Comput. **24**(2), 619–644 (2002)
36. Matthies, H.G., Keese, A.: Galerkin methods for linear and nonlinear elliptic stochastic partial differential equations. Comput. Methods Appl. Mech. Eng. **194**(12), 1295–1331 (2005)
37. Rosseel, E., De Gersem, H., Vandewalle, S.: Nonlinear stochastic galerkin and collocation methods: application to a ferromagnetic cylinder rotating at high speed. Commun. Comput. Phys. **8**, 947–975 (2010)
38. Nobile, F., Tempone, R., Webster, C.G.: A sparse grid stochastic collocation method for partial differential equations with random input data. SIAM J. Numer. Anal. **46**(5), 2309–2345 (2008)
39. Nobile, F., Tempone, R., Webster, C.G.: An anisotropic sparse grid stochastic collocation method for partial differential equations with random input data. SIAM J. Numer. Anal. **46**(5), 2411–2442 (2008)
40. Cohen, A., Devore, R., Schwab, C.: Analytic regularity and polynomial approximation of parametric and stochastic elliptic PDE's. Anal. Appl. **9**(01), 11–47 (2011)
41. Jack, A.G., Mecrow, B.C.: Methods for magnetically nonlinear problems involving significant hysteresis and eddy currents. IEEE Trans. Magn. **26**(2), 424–429 (1990)
42. Bachinger, F., Langer, U., Schöberl, J.: Numerical analysis of nonlinear multiharmonic eddy current problems. Numer. Math. **100**(4), 593–616 (2005)
43. Babuška, I., Nobile, F., Tempone, R.: Worst case scenario analysis for elliptic problems with uncertainty. Numer. Math. **101**(2), 185–219 (2005)
44. Hlaváček, I., Chleboun, J., Babuška, I.: Uncertain Input Data Problems and the Worst Scenario Method. Elsevier (2004)
45. Ben-Tal, A., Nemirovski, A.: Robust convex optimization. Math. Oper. Res. **23**(4), 769–805 (1998)

Chapter 6
Uncertainty Quantification for Magnets

In this chapter, tools and techniques developed so far will be applied to the simulation of accelerator magnets with uncertainties. Uncertainty quantification in this context is motivated by rather high accuracy requirements and considerable sensitivities with respect to manufacturing imperfections. Field distributions generated by magnets are used to guide and focus the particle beam along its trajectory in linear accelerators or storage rings. Deviations from design field distributions will cause deviations of the particles from ideal trajectories and therefore need to be kept minimal. This also explains the early and rather frequent attempts to tolerance analysis in this area. Analytical expressions for specific geometric deviations where derived in [1, 2], whereas Monte Carlo simulations were employed in [3–5]. In this work emphasis is put on a rather general approach to uncertainty quantification with a detailed exposition and analysis of the numerical methods involved. The aims are both a gain in efficiency and a better understanding of modeling uncertainty in this context. In this spirit polynomial chaos techniques have been employed in [6] to quantify variability in the context of a stochastic Brauer model for the $B - H$ curve. Numerical results reported therein will be accompanied with data for a more general stochastic material coefficient and results for the time transient case in Example 6.1. Shape uncertainties will be quantified for a model (two-dimensional) quadrupole magnet, Example 6.2, in the context of isogeometric analysis. Finally, a combination of all kinds of uncertainties as described in Sect. 4.2 will be considered for an initial design of the S-DALINAC separation dipole. For all magnets considered here, the field distribution is dominated by the iron yoke, i.e., they can be classified as iron dominated magnets, or superferric [5, p.11] in the case of the SIS-100 dipole. In contrary, coil-dominated superconducting magnets would demand for a more careful modeling and analysis of the coil shapes. Besides the aforementioned examples, field equations for magnets, appearing as a special form of the general model problem presented in Chap. 3 will be recalled in Sect. 6.1. Additionally, in this section, the so called multipole coefficients, i.e., Fourier harmonics of the magnetic field will

U. Römer, *Numerical Approximation of the Magnetoquasistatic Model with Uncertainties*, Springer Theses, DOI 10.1007/978-3-319-41294-8_6

be described. Multipole coefficients are a key quantity of interest in magnet design. A next level in design complexity would require full particle tracking simulations, which is not adapted here.

6.1 Field Equations for Magnets and Multipole Coefficients

The field distribution in magnets can be accurately described by the magnetoquasistatic model (2.24). In a preliminary design stage, as well as for magnets with small variations in the profile transverse to the beam direction and a large longitudinal to transverse dimension ratio, a two-dimensional analysis is sufficiently accurate. Depending on the operation mode, static or transient simulations are required. Magnet ramping demands for a transient analysis and will be considered for the SIS-100 dipole, Example 6.1. Note, however, that we do not consider eddy currents induced in the beam tube in this context. This constitutes a considerable challenge due to the small beam tube diameter and dedicated numerical schemes need to be designed, see [7]. In contrary, the operation mode of the S-DALINAC dipole magnets allows for a static analysis, that will be also adapted for the quadrupole magnet in Example 6.2. Another simplification consists in neglecting anisotropic effects that arise through a lamination of the iron core with the aim to suppress eddy currents [8]. Moreover, we recall that effects of hysteresis and spatial inhomogeneities of the material are disregarded here, for simplicity (Fig. 6.1).

In a next step the principles of magnetic field multipole expansions will be recalled. There exists a vast literature on this subject and we refer to [5, 9, 10]. In a two dimensional setup, as depicted, e.g., in Fig. 6.3, the beam envelope is contained in a sub-region of D_{E} marked by the dotted line with radius r_0. This is also referred to as good field region (the observer region in the context of Sect. 4.1).[1] Magnet designs aim at achieving a uniform field distribution inside this area and multipole coefficients are a suitable measure for its quantitative description. The component of the magnetic vector potential transverse to the plane, denoted u, is subject to the Cauchy-Riemann equations inside D_{E} [10], justifying the expansion

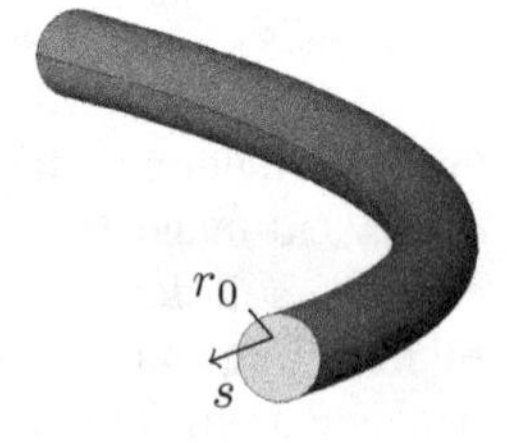

Fig. 6.1 Ideal beam trajectory in a pure dipole field and good field region within radius r_0

[1] In reality the SIS-100 beam pipe has an elliptic shape, however, we assume a circular contour here for simplicity.

$$u(r_0, \varphi) = \sum_{n=1}^{\infty} (\mathcal{F}_n(r_0) \cos(n\varphi) + \mathcal{E}_n(r_0) \sin(n\varphi)) \tag{6.1}$$

at the reference radius r_0, in local polar coordinates. The coefficients $\mathcal{F}_n, \mathcal{E}_n$ at an arbitrary radius r can be obtained by simple scaling laws [5, p.243]. A similar expansion holds true for the radial (and azimuthal) component of the magnetic flux density, i.e.,

$$B_r(r_0, \varphi) = \sum_{n=1}^{\infty} (B_n(r_0) \cos(n\varphi) + A_n(r_0) \sin(n\varphi)), \tag{6.2}$$

where the coefficients are related through

$$B_n(r_0) = \frac{-n\mathcal{F}_n}{r_0}, \quad A_n(r_0) = \frac{n\mathcal{E}_n}{r_0}, \tag{6.3}$$

see again [5]. The coefficients B_n and A_n are referred to as normal and skew multipole coefficients, respectively, and are key quantities of interest in magnet design. In particular they are at the core of many beam dynamics simulations, see also Fig. 6.2 for an illustration of several multipole field distributions. A further advantage of using multipole coefficients lies in the fact, that they are directly accessible to measurements through rotating coils [5, p.245]. Note, that depending on the given setup, other types of magnetic field representation, such as a field map, or a projection to another orthogonal system of functions might be appropriate. An ideal normal dipole magnet would be described by $B_1 \neq 0$ and $B_n = 0, n \neq 1$, whereas an ideal normal quadrupole magnet is characterized by $B_2 \neq 0$, $B_n = 0, n \neq 2$. In real magnets, the higher order harmonics B_n do not fully vanish and a good design aims at keeping them small compared to the main coefficient B_N. Modern design requirements demand for a ratio $B_n/B_N \leq 10^{-4}$ [5]. Note that, depending on symmetries, several harmonics can never occur, e.g., the nonzero multipole coefficients for the dipole and quadrupole magnet are given by

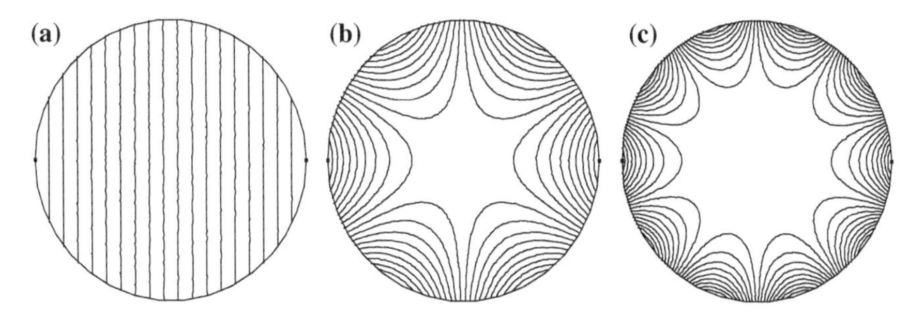

Fig. 6.2 Magnetic flux lines of the circular multipole field harmonics. **a** dipole ($n = 1$), **b** sextupole ($n = 3$), **c** decapole ($n = 5$)

$$B_1,\ B_3,\ B_5,\ B_7, \dots, \tag{6.4}$$

$$B_2,\ B_6,\ B_{10},\ B_{14}\dots \tag{6.5}$$

respectively. This might no longer hold true if symmetry is lost due to random perturbations.

In the spirit of this work, the B_n or A_n are regarded as QoI (4.2), i.e., linear functionals from the solution space into the real numbers. Indeed, based on the orthogonality relations for trigonometric functions there holds

$$B_n(r_0) = \frac{1}{\pi}\int_0^{2\pi} u(r_0, \varphi)\cos(n\varphi)\mathrm{d}\varphi. \tag{6.6}$$

It can be used to directly evaluate B_n from the finite element solution. However, we emphasize that in the context of adjoint sensitivity analysis, this would give rise to a singular right-hand side corresponding to a line current. Instead we employ the divergence theorem to derive the equivalent representation

$$B_n(r_0) = \frac{1}{r_0\pi}\int_{r=0}^{r_0}\int_0^{2\pi} (u(r,\varphi) + r\partial_r u(r,\varphi))\cos(n\varphi)\mathrm{d}\varphi\mathrm{d}r \tag{6.7}$$

which can be easily converted to the standard form (4.2). It can be shown by classical duality arguments, that the error of the finite element approximation $B_{n,h}$ of B_n, obtained by replacing u in (6.6) and (6.7) by means of u_h, converges twice as fast as the error in the energy norm. Thus we have $|B_n - B_{n,h}| = \mathcal{O}(h^2)$ in the case of lowest order finite elements and full elliptic regularity. Nevertheless, the computational cost to obtain numerical accuracies in the order of 10^{-4} and below might be high and dedicated software should be used. Higher order schemes are a promising tool to this end and we further mention the FEM-BEM coupling as described in [11] and the hybrid spectral-element finite-element scheme described in [12].

The exposition so far was restricted to two-dimensional multipole coefficients. These techniques do not apply anymore at the end of the magnet aperture where so-called fringe fields occur. In particular in a three dimensional setting, the vector potential in a cutting plane, perpendicular to the ideal trajectory, see Fig. 6.1 does not solve the Cauchy-Riemann equations anymore. However, in this case we can carry out a harmonic expansion of the integrated fields along the ideal trajectory, see Fig. 6.1. This gives rise to the definition of integrated multipole coefficients

$$\bar{B}_n(r_0) = \int_{-\infty}^{\infty} B_n(r_0, s)\mathrm{d}s, \quad \bar{A}_n(r_0) = \int_{-\infty}^{\infty} A_n(r_0, s)\mathrm{d}s, \tag{6.8}$$

where the B_n, A_n now additionally depend on the longitudinal coordinate of the trajectory. A proof that the integrated potential solves Laplace's equation can be found in [5, pp.256–257]. In practice, fields rapidly decay outside the magnet aperture and integration can be carried out along a finite curve, see also Fig. 6.1, where the ideal circular trajectory for a dipole magnet is depicted together with the good field region. Here, the integrals (6.7) and (6.8) are approximated by means of numerical quadrature.

6.2 Numerical Examples

Having established the key notions for the simulation of magnets, three detailed examples will be given to illustrate uncertainty modeling and uncertainty quantification, as described in Chaps. 4 and 5.

Example 6.1 A two-dimensional model of the SIS-100 dipole magnet [13–15] from GSI Helmholtzzentrum fr Schwerionenforschung (Darmstadt, Germany) is considered here as a numerical benchmark for the influence of uncertainties in the nonlinear material relation. This has already been investigated in [6]. There, the model of Brauer, a closed form representation for the magnetic reluctivity was fitted to randomly perturbed measured data and uncertainty propagation was carried out using a spectral projection based collocation approach. For input uncertainties of 10 % considerable variability of the multipole coefficients was observed, e.g., a ratio of approximately 1.57 of standard deviation to mean for the sextupole component. In a second step, by computing the Sobol coefficients, it was observed that the dimension of the stochastic model could be reduced from three to one or two input parameters. We complement these results by considering a different closed form model for the reluctivity

$$\nu(B) = c_4 + \frac{c_3 B^{c_2}}{c_1^{c_2} + B^{c_2}}, \tag{6.9}$$

see (4.8), together with the Karhunen-Loève expansion presented in Chap. 5. The associated eigenvalue decay will allow for a similar dimension reduction as observed in [6]. Additionally, we compare gPC and Monte Carlo techniques to the perturbation method, both for the purpose of verification and to justify the use of linearized models in this context.

An intermediate design geometry of the dipole is depicted in Fig. 6.3 on the left side. Homogeneous Dirichlet boundary conditions are applied on the entire boundary Γ_D, except for $y = 0$ where homogeneous Neumann boundary conditions are applied to model symmetry. Both the eight circular solid conductors, with an applied current of $I_0 = 7000$ A per conductor, and the iron yoke D_C include air parts. The conductivity is set to zero to account for lamination. Multipole coefficients are evaluated at a reference contour $r_0 = 25 \times 10^{-3}$ m. For the geometry modeling as well as for pre- and postprocessing the software FEMM has been used [16]. The simulation itself, consisting of a lowest order nodal finite element approximation of (3.20), has been carried out with the in-house MATLAB code NIOBE based on a FEMM triangulation. It consists of 88433 nodes with mesh refinement especially in the observer region. Linearization is carried out by means of Newton's method (3.53)

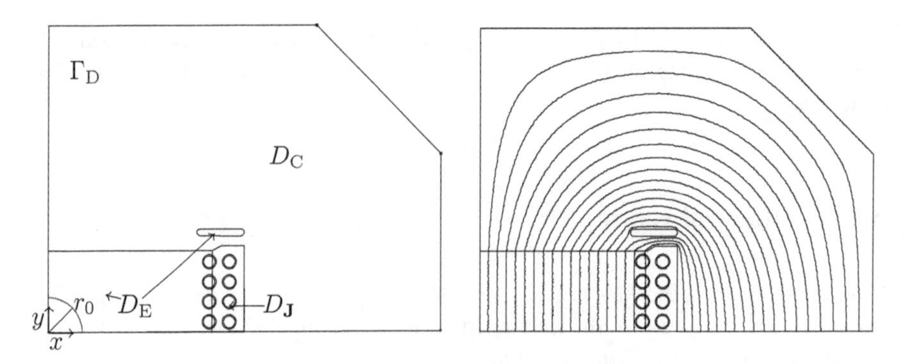

Fig. 6.3 Figure after [6]. Geometry of the SIS-100 dipole magnet on the *left*. Air inclusions in both the iron yoke and the circular conductors. Magnetic flux density lines on the *right*

and Algorithm A.1 (Appendix) with damping and the iteration process is stopped when the relative solution increment $\|u^{k+1} - u^k\|$ in the discrete L^2-norm is smaller than 1×10^{-3}.

Synthetic data for a stochastic modeling is generated as follows. Relation (6.9) is fitted to the measured data [13] by a least-square fit. The corresponding coefficients are $c_1 = 3.06$, $c_2 = 11.40$, $c_3 = 1.71 \times 10^6$ and $c_4 = 465.86$, respectively. Then each coefficient is modeled as a random variable $c_i(1 + sY_i)$, where $Y_i \sim \mathcal{U}(-\sqrt{3}, \sqrt{3})$ and $s = 0.05$. We emphasize that the parameters would typically be correlated if real data was used [17] and therefore, this modeling should be considered as a numerical benchmark, solely. This stochastic closed-form model could be readily employed in simulations as the parameters are independent. However, by solving the associated Karhunen-Loève eigenvalue problem we observe that a reduction of input random variables is possible. With the sampled covariance as an input we employ the B-spline space $\mathcal{S}_{30}^{3,1}$ on the interval $B = [0.01, 2.25]\,\mathrm{T}$ given by the initial data range. Extrapolation to all positive real numbers is carried out by the technique described in [18]. Figure 6.4 shows the first two Karhunen-Loève eigenfunctions on the right, whereas the normalized eigenvalues are $1, 1.38 \times 10^{-3}, 1.99 \times 10^{-6}, \ldots$. A widely used truncation criterion is to chose M, such that $\Psi_M > 0.95$, see (5.14), which is satisfied for our choice $M = 2$ (note that $M = 1$ would already have been sufficient). The associated random variables are determined by (5.5), however, here we simply model them to be distributed as $Y \sim \mathcal{U}(-\sqrt{3}, \sqrt{3})$. Sample trajectories of the reduced random field are depicted in Fig. 6.4 together with the initial data.

We apply the stochastic model to a time transient simulation of the magnet ramping as described in [19]. The time interval and the time step are set to $I_T = [0, 1]\mathrm{s}$ and $\delta t = 0.02\,\mathrm{s}$, respectively. Discretization is carried out by the implicit Euler method (3.86). The simplified ramping considered here, is described through the current excitation

$$I(t) = 2I_0 \begin{cases} t, & t \leq 0.5\,\mathrm{s}, \\ (1-t), & t > 0.5\,\mathrm{s}. \end{cases} \tag{6.10}$$

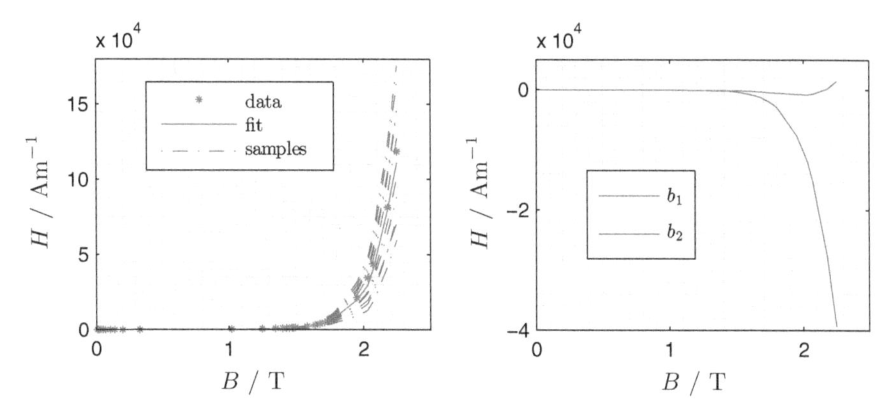

Fig. 6.4 Stochastic material model for the SIS-100 dipole magnet. Measured data [13], least-square fit by model (6.9) and ten random realizations by the Karhunen-Loève expansion with $M = 2$ on the *left*. First two Karhunen-Loève eigenfunctions on the *right*

We propagate uncertainties by the gPC collocation technique with an isotropic grid of degree two as described in Chap. 5. Figure 6.5 depicts the expected value as well as the expected value plus/minus the standard deviation for both the sextupole and the decapole coefficient. The results confirm the sensitivity of both multipole coefficients with respect to variations in the material law. For the same stochastic model, uncertainties are propagated in the static case using classical Monte Carlo sampling with 1000 samples, gPC collocation on a isotropic tensor grid of degree four and the non-intrusive perturbation technique based on finite differences. Results are reported in Tables 6.1 and 6.2 for the expected value and standard deviation, respectively. As the results of MC and gPC agree very well, these values can be considered as a reference for the perturbation method. Concerning the results for the Taylor approximation we

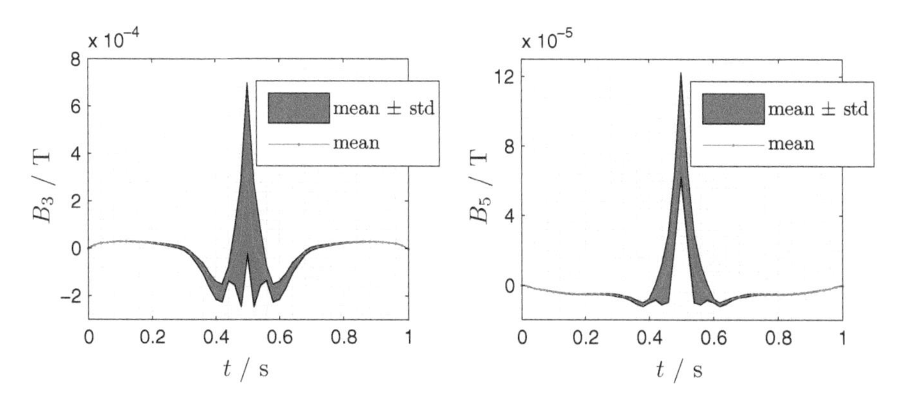

Fig. 6.5 Transient behavior of sextupole and decapole component during ramping as described by (6.10) for a gPC simulation of degree two

Table 6.1 Expected value of multipole coefficients for generalized polynomial chaos (gPC), Monte Carlo (MC) and perturbation method (per)

Multipole coefficient	MC	gPC	per
B_1	2.07	2.07	2.07
B_3	3.51×10^{-4}	3.56×10^{-4}	3.50×10^{-4}
B_5	8.84×10^{-5}	8.90×10^{-5}	9.14×10^{-5}
B_7	2.45×10^{-5}	2.46×10^{-5}	2.52×10^{-5}
B_9	2.05×10^{-6}	2.05×10^{-6}	2.09×10^{-6}

Table 6.2 Standard deviation of multipole coefficients for generalized polynomial chaos (gPC), Monte Carlo (MC) and perturbation method (per)

Multipole coefficient	MC	gPC	per
B_1	1.47×10^{-2}	1.47×10^{-2}	1.42×10^{-2}
B_3	3.60×10^{-4}	3.61×10^{-4}	3.73×10^{-4}
B_5	2.91×10^{-5}	2.89×10^{-5}	2.84×10^{-5}
B_7	3.30×10^{-6}	3.22×10^{-6}	2.70×10^{-6}
B_9	3.05×10^{-7}	3.01×10^{-7}	2.93×10^{-7}

do not observe significant variation. This indicates that the use of perturbation methods is well justified for the given example. Also the high sensitivity of the sextupole coefficient with a ratio of 1.03 of standard deviation to mean is observed confirming the results in [6]. Here, the differences in the moments of the multipole coefficients clearly originate in the modified stochastic modeling.

Example 6.2 (Quadrupole Magnet) This example is devoted to illustrate the treatment of shape uncertainties in the context of magnet design. On the left side of Fig. 6.6 the two-dimensional geometry of a model quadrupole magnet adapted from Fig. 7.1

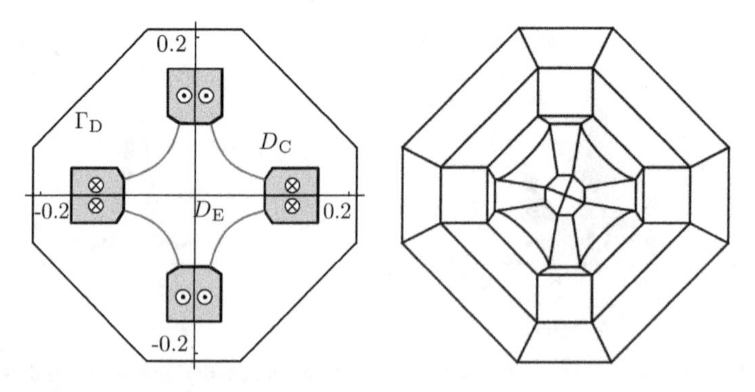

Fig. 6.6 Quadrupole geometry as given in Fig. 7.1 of [5] on the *left*. The *red* pole tips are modeled as interfaces subject to uncertainty. On the *right*, discretization in terms of multipatch NURBS. All units in meter

of [5] is depicted, where the pole tips (red colors) are subject to uncertainty. The initial pole shape is described by hyperbolas, i.e., through the relation $x/a^2 - y/b^2 = 1$ in local coordinates, where for the initial shape we have set $a = 0.05$ and $b = 0.056$, respectively. In a stochastic setting, uncertain interfaces can be modeled by choosing a or b as random variables, or equivalently by using any other CAD standard shape representation with random parameters, see the discussion in Sect. 4.2.4, in particular Fig. 4.2. Then, again a truncated Karhunen-Loève expansion can be used to obtain a reduced number of uncorrelated random inputs. Here, we adapt the isogeometric analysis methodology as outlined in Example 3.3.2. A multi-patch configuration, consisting of 36 patches is chosen to discretize the geometry as depicted in Fig. 6.6 on the right side. Each patch is described by the NURBS mapping

$$\mathbf{F}(\xi, \eta) = \sum_{i=1,\ldots,3} \sum_{j=1,\ldots,3} R_{ij}(\xi, \eta)\mathbf{c}_{ij}, \tag{6.11}$$

where R_{ij} are second degree NURBS basis functions with $\mathcal{C}^1$-continuity. In particular, this allows for an exact representation of hyperbolic shapes. The patch interfaces are matched in the standard conforming way, i.e., the underlying knot vectors are constructed identically, yielding a global $\mathcal{C}^0$ parametrization. Focusing on shape variations we model the material to be linear with a constant permeability $\mu = 4\pi 10^{-2}$ H/m. A total piecewise constant current of 15 MA is supplied for each of the four conductor parts of Fig. 6.6. Homogeneous Dirichlet boundary conditions are applied on the whole boundary Γ_D. Following the iso-parametric concept a basis for $\mathcal{H}_h(\mathbf{grad}, D)$ is constructed in the same space as the mapping $\mathbf{F}$ and we refer to Sect. 3.3.2 for details. The multipole coefficients are evaluated at a reference radius of $r_0 = 20$ mm according to (6.7). For simplicity, the solution is interpolated at the Gauss integration points used in the matrix assembly solely, and therefore, to increase accuracy a quadrature of degree five is employed.

We focus on the uncertainty in the quadrupole gradient, defined as $g := 2B_2/R_0^2$. As we are concerned with several parameters and only one cost function, adjoint techniques as described in Sect. 4 are well suited to this end. Here, no assumptions, despite the $\mathcal{C}^1$-smoothness, are made for the shape uncertainty and, therefore, we resort to a worst-case analysis. The maximum deviation of g from its design value $g_0 = 19.73\,\mathrm{T/m^2}$ is investigated for different levels of shape parametrization, characterized by the number of free control points. On the coarsest level the hyperbola is described by three control points, however, the end points are kept fix to avoid variability in the singular points, as this would require a more general shape calculus as presented here. Hence, only one control point per pole is subject to uncertainty. Through knot refinement this number is increased by means of one on each level, up to level four. After refining the geometry, in each direction every mesh cells is divided by twenty. For an evaluation of the accuracy of the numerical approximation of the quadrupole gradient g, see Table 6.3. There, the error on each level is estimated with respect to a fine discretization consisting of 367641 total Degrees Of Freedom (DOF), denoteds $\Delta_h g$. Additionally, the numerical computation of the shape gradient as stated in Proposition 4.12 is verified. To this end in Table 6.3 the

Table 6.3 Quadrupole discretization error and gradient verification at different mesh levels. Error estimate of the shape gradient, evaluated as in Proposition 4.12 and finite difference approximation (5.65) for different number of perturbed control points. The reference for g is computed with 367641 DOF

Parameter	Level 1	Level 2	Level 3	Level 4
$\Delta_h g/g_0$ (%)	0.56	0.05	<0.01	<0.01
$\Delta_{\mathrm{FD}}\delta g$	5.85×10^{-5}	3.96×10^{-5}	7.29×10^{-5}	1.38×10^{-4}

Table 6.4 Worst case scenario by linearization and error estimated by means of MATLAB's fmincon routine as outlined in Sect. 5.2.5. Perturbation magnitude of $s = 0.1$

Num. free CP	$\mathrm{wcs}_{\mathrm{L}}(g)/g_0$	$\mathrm{wcs}^*(g)/g_0$	Rel. diff. (%)
4	7.82	8.48	7.78
8	16.77	19.77	15.17
12	18.57	21.77	14.70

maximum deviation with respect to a finite difference gradient computation (5.65), denoted Δ_{FD}, is given. Both estimated errors are found to be sufficiently small. We emphasize that no correlation is imposed here. If knowledge of the shape perturbations were available, e.g., in terms of measurements, the correlation structure could be incorporated by means of convex constraints in a worst-case scenario context as outlined in [20, p.10] and described in Sect. 5.2.5. In Table 6.4 numerical results for the different parametrization levels are presented. Not surprisingly, a significantly smaller worst-case estimate is obtained for the coarse parametrization. In this case, large perturbations in the control point are necessary to obtain a comparable shape perturbation to the finer parametrization levels. We again refer to Fig. 4.2 for an illustration. We compute the worst-case scenario by a Taylor expansion $\mathrm{wcs}_{\mathrm{L}}(g)$ and by directly solving the optimization problem (5.83), denoted $\mathrm{wcs}^*(g)$. Here, for the latter case MATLAB's fmincon routine is used to carry out sequential quadratic programming. We observe a difference of about 15 % and infer, that the problem is rather sensitive to shape perturbations and first order approximations should be used to obtain rough estimates of the output uncertainties, solely.

Example 6.3 (S-DALINAC Separation Dipole Magnet) The last example of this section is concerned with the new design of the separation dipole of the Superconducting-DArmstadt-LINear-ACcelerator (S-DALINAC). The S-DALINAC is a linear, recirculating accelerator, i.e., the particle beam is redirected several times to the same acceleration part. Each recirculation follows a different path with different energies, and the beam is assigned to the appropriate path of recirculation by a so-called separation dipole. Figure 6.7 on the left depicts an intermediate design with four ideal trajectories (blue) and the respective good field regions, corresponding to four different energy levels. The magnet was designed by the Institut für Kernphysik of Technische Universität Darmstadt and the company Sigmaphi. Simulations during the design phase were carried out in collaboration with the Institut für Kernphysik.

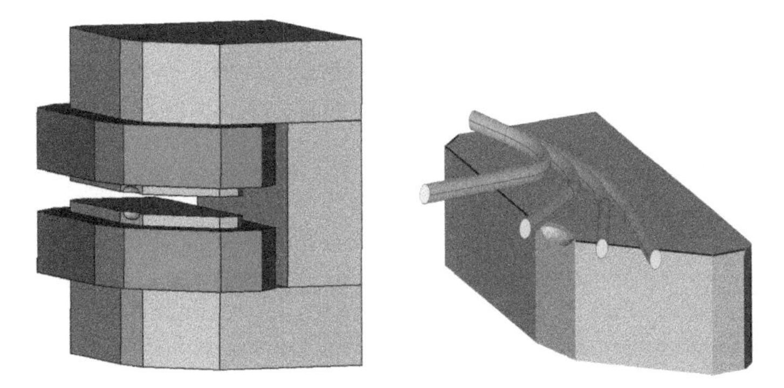

Fig. 6.7 Simplified, intermediate design of the S-DALINAC separation dipole magnet without mirror plates and with ideal trajectories in *blue* on the *left*. Iron yoke in *grey* and coils in *red color*. Designed by the Institut für Kernphysik of TU Darmstadt and Sigmaphi. Iron yoke with chamfer (*red line*) on the *right*

Mirror plates, focusing the dipole field to the aperture are not shown here. Coil parts are depicted in red color and modeled, according to the stranded conductor model, with 60 turns, whereas the grey parts represent the iron yoke. In this case, the shape of the yoke impedes a two-dimensional analysis, however, as the magnets are not ramped, static simulations can be carried out. One edge of the yoke is chamfered as depicted in Fig. 6.7 on the right in order to decrease undesired higher order multipole coefficients. We cannot give a detailed description of chamfering here and instead refer to [9, p.254]. Roughly speaking, the main idea consist in compensating for the inhomogeneous integrated magnetic field along the edges at the end of yoke. As the maximum is attained in the middle of the edge removing iron at this part is a suitable means. Here, we consider integrated multipole coefficients (6.8), in particular the dipole component, for the second beam passing the chamfered egde in red. The chamfer profile is modeled by a spline with two uniformly distributed control points. Nonlinear magnetostatic simulations are carried out with the second order tetrahedral solver of CST EM STUDIO®. A solver accuracy of 1×10^{-4} is chosen and the maximum step size in the good field region is set to 2 mm (the maximum yoke dimension is in the order of 300 mm). Adaptive energy based local refinement is carried out until the relative change is smaller than 5×10^{-4}, with a maximum number of eight refinement cycles. The nonlinear reluctivity is again modeled by (6.9) and interpolated to be used within CST EM STUDIO®. Finally, a maximum number of 25 nonlinear iteration steps are carried out with a nonlinear accuracy of 1×10^{-3}.

All three different kinds of inputs are considered here and modeled as follows. Let Y_{J} refer to the coil current I with an initial value of $I_0 = 132\,\mathrm{A}$. For the perturbation interval we chose $\Gamma_{\mathrm{J}} = Y_{\mathrm{J}}(1 \pm 0.005)$, which is already large compared to the technical specifications. Concerning shape parameters, as already mentioned, we choose the two control points c_1, c_2 of the spline chamfering, describing the chamfer shape perpendicular to the edge. We choose local coordinates, centered at the beginning of the chamfer near the left edge point, with an x-coordinate in parallel to the edge of length l. Then we have $\mathbf{Y}_{\Gamma_\mathrm{I}} = (c_{1,y}, c_{2,y})$ with initial value $\mathbf{c}_1 = (l/3, 10)$, $\mathbf{c}_2 = (2l/3, 10)$. Here, the chosen perturbation interval of $\Gamma_{\Gamma_\mathrm{I}} = \mathbf{Y}_{\Gamma_\mathrm{I}}(1 \pm 0.5)$ corresponds to maximum deviations in the order of two millimeters. This is a reasonable assumption for the corresponding manufacturing process. Finally, for the material law with initial values $\mathbf{c} = (4.28 \times 10^5, 1.24 \times 10^6, 7.16 \times 10^3, 1.26 \times 10^2)$, we chose $\mathbf{Y}_\nu = (c_1, c_2)$ and a perturbation interval of $\Gamma_\nu = \mathbf{Y}_\nu(1 \pm 0.2)$. This is a modeling assumption in absence of measurements for the material under consideration. The combined input vector is then given by $\mathbf{Y} = (\mathbf{Y}_\mathbf{J}, \mathbf{Y}_{\Gamma_\mathrm{I}}, \mathbf{Y}_\nu)$ and $\Gamma = \Gamma_\mathbf{J} \times \Gamma_{\Gamma_\mathrm{I}} \times \Gamma_\nu$. We note that $M = 5$ which is already challenging in an uncertainty quantification context, in view of the model's complexity. Hence, instead of running Monte Carlo simulations or constructing tensor product grids we aim at a reduction of the input dimension. To this end a cut-HDMR expansion as described in Sect. 5.2 is carried out. The result is reported in Fig. 6.8. The finite difference approximations have been obtained with a step width of $1/10$ of the respective interval of Γ. The results indicated that combined second order effects are the most significant in particular if coil current perturbations are involved. Variations in both current and reluctivity have more significant effects than shape variations, however, one should keep in mind, that the perturbation intervals are possibly over estimated in these cases. For comparison we carry out a gPC simulation of degree two with uniformly distributed parameters for the standard deviation of Y_1 and Y_{14}, separately. In particular this serves at validating whether the cut-HDMR estimates of the importance of different order contributions are justified. Given cut-HDMR importance values of the same order of magnitude we expect that a similar amount of standard deviation can be ascribed to both. As we obtain $\mathrm{Std}_1 = 0.2831$ Tmm and $\mathrm{Std}_{14} = 0.2834$ Tmm, respectively, this can be confirmed. Hence, the cut-HDMR expansion provides a suitable means for model

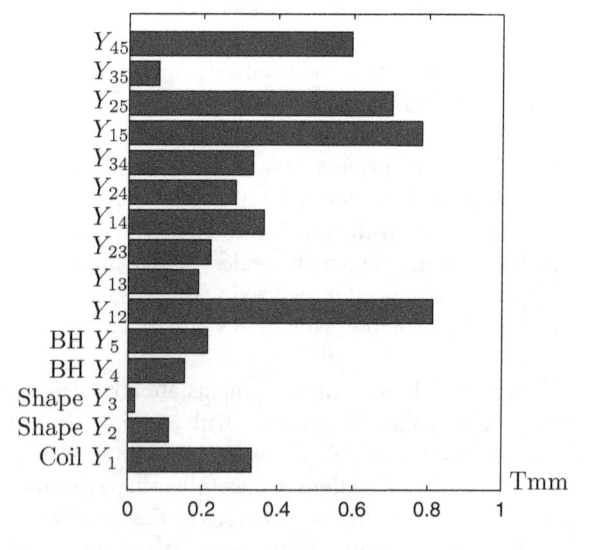

Fig. 6.8 Relative importance of uni- and bivariate contributions by means of a second order cut-HDMR expansion for the integrated dipole component of the S-DALINAC separation dipole magnet. The nominal value for the integrated dipole is 1.87×10^2 Tmm

reduction as unimportant inputs can be identified and removed. Moreover, the dominant combined effects in this case, could not have been computed by a first order perturbation method.

References

1. Halbach, K.: First order perturbation effects in iron-dominated two-dimensional symmetrical multipoles. Nuclear Instrum. Methods **74**(1), 147–164 (1969)
2. Halbach, K.: Fields and first order perturbation effects in two-dimensional conductor dominated magnets. Nuclear Instrum. Methods **78**(2), 185–198 (1970)
3. Scandale, W., Todesco, E., Tropea, P.: Influence of mechanical tolerances on field quality in the lhc main dipoles. IEEE Trans. Appl. Supercond. **10**(1), 73–76 (2000)
4. Ferracin, P., Scandale, W., Todesco, E., Wolf, R.: Modeling of random geometric errors in superconducting magnets with applications to the cern large hadron collider. Phys. Rev. Special Topics-Accelerators Beams **3**(12), 122403 (2000)
5. Russenschuck, S.: Field Computation for Accelerator Magnets: Analytical and Numerical Methods for Electromagnetic Design and Optimization. Wiley (2011)
6. Bartel, A., De Gersem, H., Hülsmann, T., Römer, U., Schöps, S., Weiland, T.: Quantification of uncertainty in the field quality of magnets originating from material measurements. IEEE Trans. Magn. **49**, 2367–2370 (2013)
7. Koch, S., Trommler, J., De Gersem, H., Weiland, T.: Modeling thin conductive sheets using shell elements in magnetoquasistatic field simulations. IEEE Trans. Magn. **45**(3), 1292–1295 (2009)
8. Koch, S., De Gersem, H., Weiland, T., Fischer, E., Moritz, G.: Transient 3d finite element simulations of the sis100 magnet considering anisotropic, nonlinear material models for the ferromagnetic yoke. IEEE Trans. Appl. Supercond. **18**(2), 1601–1604 (2008)
9. Tanabe, J.T.: Iron Dominated Electromagnets: Design, Fabrication, Assembly and Measurements. World Scientific, Singapore (2005)
10. Rossbach, J., Schmüser, P.: Basic course on accelerator optics. CERN European Organization for Nuclear Research-Reports-CERN, pp. 17–17 (1994)
11. Auchmann, B.: The Coupling of Discrete Electromagnetism and the Boundary-Element Method for the Simulation of Accelerator Magnets. PhD thesis, Vienna (2004)
12. De Gersem, H.: Combined spectral-element, finite-element discretization for magnetic-brake simulation. IEEE Trans. Magn. **46**(8), 3520–3523 (2010)
13. Kovalenko, A.D., Kalimov, A., Khodzhibagiyan, H.G., Moritz, G., Muhle, C.: Optimization of a superferric nuclotron type dipole for the gsi fast pulsed synchrotron. IEEE Trans. Appl. Supercond. **12**(1), 161–165 (2002)
14. Agapov, N., Averichev, S., Donyagin, A., Eliseeva, I., Karpunina, I., Khodzhibagiyan, H., Kovalenko, A., Kuznetsov, G., Kuzichev, V., Moritz, G., et al.: Experimental study of a prototype dipole magnet with iron at t= 80 k for the gsi fast cycling synchrotron. IEEE Trans. Appl. Supercond. **12**(1), 116–119 (2002)
15. Koch, S., De Gersem, H., Weiland, T.: Transient 3d finite element simulations of the field quality in the aperture of the sis-100 dipole magnet. IEEE Trans. Appl. Supercond. **19**(3), 1162–1166 (2009)
16. Meeker, D.: Finite element method magnetics. Version 4.2 (1 April 2009 Build) (2010)
17. Ramarotafika, R., Benabou, A., Clénet, S.: Stochastic modeling of soft magnetic properties of electrical steels, application to stators of electrical machines. IEEE Trans. Magn. **48**, 2573–2584 (2012)
18. Reitzinger, S., Kaltenbacher, B., Kaltenbacher, M.: A note on the approximation of B-H curves for nonlinear computations. Technical Report 02-30, SFB F013, Johannes Kepler University Linz, Austria (2002)

19. De Gersem, H., Koch, S., Shim, S.Y., Fischer, E., Moritz, G., Weiland, T.: Transient finite-element simulation of the eddy-current losses in the beam tube of the sis-100 magnet during ramping. IEEE Trans. Appl. Supercond. **18**(2), 1613–1616 (2008)
20. Babuška, I., Nobile, F., Tempone, R.: Worst case scenario analysis for elliptic problems with uncertainty. Numer. Math. **101**(2), 185–219 (2005)

Chapter 7
Conclusion and Outlook

7.1 Conclusion

In this work, uncertainty quantification for the nonlinear magnetoquasistatic model with application to magnet design has been addressed. The model was found to be well-posed, in particular with continuous dependency on the input data. Moreover derivatives with respect to the $B - H$ curve, the shape of the iron/air interface and the current excitation could be obtained.

Even in the presence of nonlinearities, first order sensitivity analysis techniques could be used to efficiently propagate uncertainties for several real-world magnet configurations. For example, differences with respect to a Monte Carlo reference were negligible for the SIS-100 dipole magnet but the solution time was significantly increased. Also for an electrical transformer, reasonable approximations of the statistical moments were found by means of a first order stochastic Taylor expansion. However, in the case of a worst-case scenario in combination with shape perturbations, the results obtained by means of linearization were not sufficiently accurate. Here, linearization errors up to 15 % were estimated for moderate input perturbations. In conclusion, error estimators should be established for perturbation methods in order to increase their reliability.

An alternative approach was given by a non-intrusive stochastic collocation method. As opposed to perturbation techniques, higher order schemes are available and the method was mathematically proven to converge faster than the classical Monte Carlo method, asymptotically. In contrast to linear elliptic models, its performance was more difficult to analyze in presence of a nonlinearity and the estimated convergence rate was algebraic.

It was argued that the input randomness of the model could be approximated by a few random variables in many cases and the Karhunen-Loève expansion has been presented as an efficient tool to this end. With respect to the $B - H$ curve and the examples considered in this work, no more than three input parameters were required. Additionally, uncorrelated parameters were obtained as opposed to correlated parameters in closed-form representations.

U. Römer, *Numerical Approximation of the Magnetoquasistatic Model with Uncertainties*, Springer Theses, DOI 10.1007/978-3-319-41294-8_7

With regard to uncertainties in magnet design, strong sensitivities have been observed for the multipole coefficients. In particular for higher order coefficients the standard deviation can be of the same order of magnitude as the mean value.

7.2 Outlook

Several interesting and important aspects of uncertainty quantification in the context of accelerator magnets could not be covered in this thesis. A heterogeneous and anisotropic modeling of material coefficients would require a more general stochastic modeling, such as a vector-valued and high-dimensional Karhunen-Loève expansion. Also the modeling of uncertainties in the shape and positioning of conductor strands was not addressed.

The number of input parameters in real-life applications can quickly become very high. For these cases adjoint sensitivity analysis techniques have been presented, as well as a reduction of the parametric-dimension, e.g., by the cut-HDMR expansion. Many other schemes for high-dimensional integration have been proposed in the literature, in particular sparse grid and low-rank tensor approximations. Their analysis and application to the models and examples considered here is the subject of future work. Moreover, an a posteriori error analysis of the stochastic error should be addressed, which is in particular important for the perturbation methods presented here. Furthermore, robust optimization should be mentioned as a promising tool to increase the reliability of a design.

For simplicity, several results were derived in static or two-dimensional settings and their extension to the three-dimensional transient case was postponed to future work. Emphasis was put on the material law as an input parameter. Here, too, results could be generalized to other types of inputs. In particular, in the context of magnet design, coil-dominated magnets have only been partially addressed here, as uncertainties in the strands would be much more important.

Appendix A
Linearization

Algorithm A.1 (*Iterative Linearization Procedure*)

(B.1) set $l = 1$, choose a relaxation parameter $\alpha \in (0, 1]$, a tolerance tol, a maximum number of iterations $l_{\max}$ and an initial value $\mathbf{A}^{\{1\}}$
(B.2) set $\mathbf{A}_{\mathrm{L},0} = \mathbf{A}^{\{l\}}$
(B.3) solve (3.51) to obtain $\mathbf{A}_{\mathrm{L}}$ and update $\mathbf{A}^{\{l+1\}} = \alpha \mathbf{A}_{\mathrm{L}} + (1 - \alpha)\mathbf{A}_{\mathrm{L},0}$
(B.4) evaluate the linearization error $\mathrm{err}_{\mathrm{L}} := \|\mathbf{A}^{\{l+1\}} - \mathbf{A}^{\{l\}}\|$ in a suitable norm
(B.5) if $\mathrm{err}_{\mathrm{L}} \leq$ tol or $l > l_{\max}$ stop, else go to (B.2) and set $l = l + 1$

For the following definition see [1].

Definition A.1 (*Q-Convergence Order*) A sequence $(x^{\{l\}})$ is said to converge linearly to x^*, if their exists a constant $C \in (0, 1)$ such that

$$\|x^{\{l+1\}} - x^*\| \leq C\|x^{\{l\}} - x^*\| \tag{A.1}$$

with a suitable norm. If in addition, the constant C can be replaced by a zero sequence $(a^{\{l\}})$ in (A.1) we speak of superlinear convergence. The sequence is said to converge quadratically if

$$\|x^{\{l+1\}} - x^*\| \leq C\|x^{\{l\}} - x^*\|^2 \tag{A.2}$$

holds.

U. Römer, *Numerical Approximation of the Magnetoquasistatic Model with Uncertainties*, Springer Theses, DOI 10.1007/978-3-319-41294-8

Appendix B
B-Splines and NURBS

B-splines have been found to provide a suitable framework for both analysis and computation and some of its main properties are briefly recalled here, see [2, 3]. On an interval $I \subset \mathbb{R}$, setting $\min(\bar{I}) = I_{\min}$, $\max(\bar{I}) = I_{\max}$ we introduce a sequence of knots

$$\tau_N := \xi_1 \leq \xi_2 \leq \cdots \leq \xi_{N+q+1}. \tag{B.1}$$

Knot multiplication up to q times is allowed and we assume that the knot vector is open, i.e., the end knots are repeated $q + 1$ times. The knot vector without repetition is referred to as Π with step sizes $\Delta_n, n = 1, \ldots, d$. B-splines are obtained following a recursive procedure. For $q = 0$ we define piecewise constant functions as

$$B_i^0(\xi) = \begin{cases} 1 & \xi_i \leq \xi \leq \xi_{i+1}, \\ 0 & \text{otherwise.} \end{cases} \tag{B.2}$$

Then for $q = 1, 2, 3, \ldots$ we set

$$B_i^q(\xi) = \frac{\xi - \xi_i}{\xi_{i+1} - \xi_i} B_i^{q-1}(\xi) + \frac{\xi_{i+p+1} - \xi}{\xi_{i+p+1} - \xi_{i+1}} B_{i+1}^{q-1}(\xi). \tag{B.3}$$

These functions span the space $\mathcal{S}_N^{q,k} = \text{span}\{B_j^q\}_{j=1}^N$ of polynomials of degree q on sub-intervals of Π and minimum regularity k at the knots. B-splines of degree q possess regularity $p - r_i$ at knot ξ_i duplicated r_i times. Varying regularity from knot to knot, is expressed by means of a vector $\mathbf{k}$, i.e., splines in $\mathcal{S}_N^{q,\mathbf{k}}$ are k_i-times continuously differentiable at the i-th knot. An example of second degree splines in

U. Römer, *Numerical Approximation of the Magnetoquasistatic Model with Uncertainties*, Springer Theses, DOI 10.1007/978-3-319-41294-8

depicted in Fig. B.1. Based on B-splines, rational basis functions as used to represent NURBS curves, faces and volumes are defined as

$$R_i^q(\xi) = \frac{B_i^q(\xi)w_i}{\sum_{i=1}^N B_i^q(\xi)w_i}, \tag{B.4}$$

where w_i are referred to as weights.

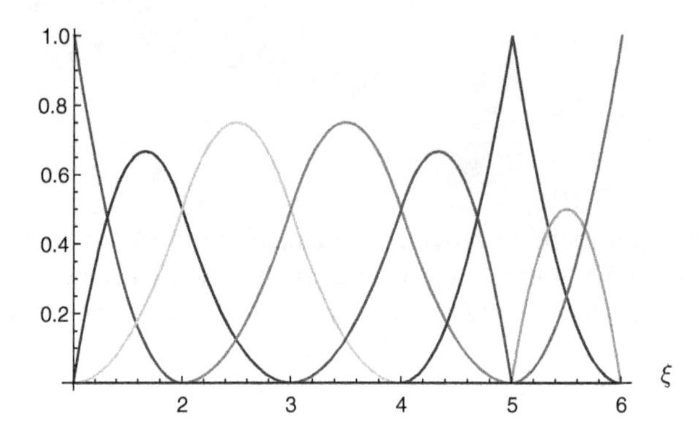

Fig. B.1 B-splines with $q = 2$ and $k = 0$ associated to the knot vector $(1, 1, 1, 2, 3, 4, 5, 5, 6, 6, 6)$

Appendix C
Higher Order Sensitivity Analysis

We recall the definition of the vector function

$$\mathbf{h}(y,\ \mathbf{curl}\,\mathbf{A}(y)) = \nu(y, |\,\mathbf{curl}\,\mathbf{A}(y)|)\ \mathbf{curl}\,\mathbf{A}(y). \tag{C.1}$$

Differentiating the magnetostatic formulation

$$\mathbf{curl}\ (\nu(y, |\,\mathbf{curl}\,\mathbf{A}(y)|)\ \mathbf{curl}\,\mathbf{A}(y)) = \mathbf{J}, \tag{C.2}$$

w.r.t. y requires derivatives $\partial_y^k \mathbf{h}(y, \mathbf{r}(y))$. By Faà di Bruno's formula we have

$$\partial_y^k \mathbf{h}(\cdot, \mathbf{r}(y)) = \sum_{\pi \in \Pi_k} D_{\mathbf{r}}^{|\pi|} \mathbf{h}(\cdot, \mathbf{r}(y)) \left(\partial_y^{|\pi_1|} \mathbf{r}(y), \ldots, \partial_y^{|\pi_{|\pi|}|} \mathbf{r}(y)\right),$$

where Π_k is the set of all partitions of $\{1, 2, \ldots, k\}$ and $|\pi|$ the number of blocks in $\pi = \{\pi_1, \ldots, \pi_{|\pi|}\}$. Note that the l-th derivative $D_{\mathbf{r}}^l \mathbf{h}(\cdot, \mathbf{r}(y))$ is considered to be a multilinear map $D_{\mathbf{r}}^l \mathbf{h} : \mathbb{R}^{3l} \to \mathbb{R}^3$ that is bounded by the Assumption 5.9. Observing, that the term associated to $|\pi| = 1$ is equal to $\boldsymbol{\nu}_{\mathrm{d}} \partial_y^k \mathbf{r}(y)$ we obtain

$$\begin{aligned}&\mathbf{curl}\ \left(\boldsymbol{\nu}_{\mathrm{d}}(y,\ \mathbf{curl}\,\mathbf{A})\ \mathbf{curl}\ \partial_y^k \mathbf{A}\right) = \\ &- \mathbf{curl}\left(\sum_{l=0}^{k} \binom{k}{l} \sum_{\substack{\pi \in \Pi_{k-l},\\ |\pi| \neq 1, l=0}} D_{\mathbf{r}}^{|\pi|} \partial_y^l \mathbf{h}(y,\ \mathbf{curl}\,\mathbf{A}) \left(\partial_y^{|\pi_1|}\ \mathbf{curl}\,\mathbf{A}, \ldots, \partial_y^{|\pi_{|\pi|}|}\ \mathbf{curl}\,\mathbf{A}\right)\right).\end{aligned} \tag{C.3}$$

We now recall Theorem 5 and give a proof.

Theorem C.1 *Let $\boldsymbol{\nu}_{\mathrm{d}}(\mathbf{y}, \cdot)$ fulfill Assumption 3.10 (ρ-) a.e. and Assumption 5.9 hold true. The collocation approximation $\mathbf{A}_q$ converges to $\mathbf{A}$, as*

U. Römer, *Numerical Approximation of the Magnetoquasistatic Model with Uncertainties*, Springer Theses, DOI 10.1007/978-3-319-41294-8

$$\|\mathbf{A} - \mathbf{A}_q\|_{L^2_\rho(\Gamma)\otimes W_{\mathrm{st}}(D)} \leq C_1 q^{-1}. \tag{C.4}$$

Additionally, the collocation error for the finite element solution $\mathbf{A}_h$ converges as

$$\|\mathbf{A}_h - \mathbf{A}_{h,q}\|_{L^2_\rho(\Gamma)\otimes W_{\mathrm{st}}(D)} \leq C_2 q^{-k}, \tag{C.5}$$

where C_2 depends on k, h.

Proof We first proof estimate (C.5). In a first step, we observe that the collocation error is an interpolation error, i.e.,

$$\mathbf{A}_h - \mathbf{A}_{h,p} = \mathbf{A}_h - \mathcal{I}_p \mathbf{A}_h. \tag{C.6}$$

Following standard arguments [4], the error is related to best approximation error in $\mathbb{Q}_p(\Gamma) \otimes W_{\mathrm{st}}(D)$ as

$$\|\mathbf{A}_h - \mathcal{I}_p \mathbf{A}_h\|_{L^2_\rho(\Gamma)\otimes W_{\mathrm{st}}(D)} \leq C \inf_{\mathbf{v}\in\mathbb{Q}_p(\Gamma)\otimes W_{\mathrm{st}}(D)} \|\mathbf{A}_h - \mathbf{v}\|_{L^\infty(\Gamma, W_{\mathrm{st}}(D))}. \tag{C.7}$$

Given a bounded derivative $\partial_y^k \mathbf{A}_h \in L^\infty(\Gamma, W_{\mathrm{st}}(D))$ a result of Jackson yields

$$\|\mathbf{A}_h - \mathcal{I}_p \mathbf{A}_h\|_{L^2_\rho(\Gamma)\otimes W_{\mathrm{st}}(D)} \leq C p^{-k} \|\partial_y^k \mathbf{A}_h\|_{L^\infty(\Gamma, W_{\mathrm{st}}(D))}, \tag{C.8}$$

cf. [5]. Hence, it remains to show that $\partial_y^k \mathbf{A}_h \in L^\infty(\Gamma, W_{\mathrm{st}}(D))$. This was established in [6, Lemma 3], using the fact that $\mathbf{A}_h$ belongs to a finite dimensional space and hence all norms are equivalent.

For estimate (C.4) we simply observe that $\mathbf{G}_1 = \partial_y \mathbf{h}(y, \mathbf{curl}\, \mathbf{A})$. Hence, $\mathbf{G}_1 \in L^2(D)^3$ and $\partial_y \mathbf{A} \in W_{\mathrm{st}}(D)$. Then (C.4) is established using again the estimate of Jackson. □

References

1. Schatzman, M.: Numerical Analysis. Clarendon Press (2002)
2. Hughes, T.J.R., Cottrell, J.A., Bazilevs, Y.: Isogeometric analysis: cad, finite elements, nurbs, exact geometry and mesh refinement. Comput. Methods Appl. Mech. Eng. **194**(39), 4135–4195 (2005)
3. De Boor, C.: A Practical Guide to Splines, vol. 27. Springer, New York (1978)
4. Babuška, I., Nobile, F., Tempone, R.: A stochastic collocation method for elliptic partial differential equations with random input data. SIAM Rev. **52**(2), 317–355 (2010)
5. Motamed, M., Nobile, F., Tempone, R.: A stochastic collocation method for the second order wave equation with a discontinuous random speed. Numer. Math. **123**(3), 493–536 (2013)
6. Römer, U., Schöps, S., Weiland, T.: Stochastic modeling and regularity of the nonlinear elliptic curl-curl equation. SIAM/ASA J Uncertainty Quantification (in press)

Curriculum Vitae

Name	Ulrich Römer
Address	Emilstr. 2, 64289 Darmstadt, Germany
Date of Birth	25.06.1983
Nationality	German
Email	roemer@temf.tu-darmstadt.de

Education

Feb. 2015– Research Assistant
Research group leader, topic: Uncertainty Quantification in Electromagnetics, Institut für Theorie Elektromagnetischer Felder, TU Darmstadt

2009–2015 Ph.D. thesis
Department of Electrical Engineering and Information Technology, TU Darmstadt, title: *Numerical Approximation of the Magnetoquasistatic Model with Uncertainties and its Application to Magnet Design*

2003–2009 Diploma thesis
Department of Electrical Engineering and Information Technology, TU Darmstadt, title: *Electric Field Singularities at Dielectric Triple Points*
Generalist Engineer program and double degree, École Centrale de Lyon, France (2005–2007)

1999–2002 High school graduation
Claus-von-Stauffenberg Schule, Rodgau, Germany

U. Römer, *Numerical Approximation of the Magnetoquasistatic Model with Uncertainties*, Springer Theses, DOI 10.1007/978-3-319-41294-8

Awards and Honors

2015 Appointment as a GAMM Junior, International Association of Applied Mathematics and Mechanics (GAMM)

2010 Pepperl+Fuchs price for the best diploma with distinction, Department of Electrical Engineering and Information Technology, TU Darmstadt

CPSIA information can be obtained
at www.ICGtesting.com
Printed in the USA
LVHW06s2059040818
585760LV00027B/17/P